Introduction By Dave Wolin

I'm calling myself the compiler as I'm not an author and someone has to take the blame, my other colleagues who helped wished to remain anonymous, (we don't blame them) since this is really a compilation - so all facts, alternative facts, opinions and remembrances are the responsibility of whoever gave them to us. Please don't complain if you disagree with the opinions or statements herein.

In compiling this book, I guess I have to state that I'm pretty qualified to discuss the Long Beach Grand Prix. Before the first race, in 1975, a client of my ad agency / p.r. / marketing firm was approached with an investment opportunity; buy some stock in the yet to happen Long Beach Grand Prix, He was a cheap b---d, said no, too risky.......... At the first race, in 1975, I met with all the teams, giving them catalogs and advice. I knew most of them from my own Formula A / 5000 experience; they didn't hold I against me.

During the Formula One years, I often teamed up with Carroll Smith and Mac Tilton as we all had something to offer and this was a rare close to home opportunity. When CART came along, again I knew most of the teams as I had managed a garage at the Indianapolis Motor Speedway for many years, providing parts and technical support. I was an early supporter of the Grand Prix Expo, exhibiting my clients products

And as the support races grew, I renewed my relationships with IMSA friends, many of whom I'd raced with

Today I operate the non profit Racing History Project (www.racinghistoryproject.com), write a few books and magazine articles, still do some racing and driver coaching and builds a new hot rod every couple of years. I have to thank my wife, Jane, for tolerating all the time I've put into this and all the other books and the invaluable help from all those who have participated at Long Beach by racing, taking photos or helping in many ways.

To quote Mark Twain, *"Persons attempting to find a motive in this narrative will be prosecuted; persons attempting to find a moral in it will be banished; persons attempting to find a plot in it will be shot."*

Dave Wolin

Long Beach, the seventh largest city in California, 358,000 people in 1970, had a staid, boring image. Known as "Iowa by the Sea" for it's midwestern transplants and buying the aging Queen Mary in 1967 didn't attract as much attention as was hoped for. The downtown area and "The Pike" amusement park, renamed Queens Park, was primarily pawn shops, tattoo parlors, sleazy hotels and porno theaters.

Racing on public roads had a long history in Southern California, from 1903 when Barney Oldfield won at Agricultural Park, where the Coliseum now stands. Wilshire Boulevard, Ocean Avenue in Santa Monica, Corona, on what's now Grand Avenue, where 150,000 people watched Eddie Pullen win.

In 1974 the city of Long Beach began a billion dollar redevelopment program of the downtown area. British expatriate Chris Pook, a Sorbonne educated, one time race car driver and owner of a local travel agency proposed to the Long Beach Convention and Visitors Bureau and Long Beach City Council that they consider holding a street race similar to the Formula One Monaco Grand Prix. Pitching his plan to the Chamber of Commerce, local Rotary Clubs, Lions Clubs, the Coastal Commission and numerous business organizations, he promised huge tourism dollars, full hotels and the elevation of Long Beach to a world class city. Conveniently, his travel agency was in the same building as the Visitors Bureau and they had a nodding relationship and the Visitors Bureau, after building a new convention center in 1972, felt overshadowed by Anaheim's entry in the convention center business..

He was able to convince the international sanctioning body of Formula One to hold, for the first time ever, a second street race, the other being Monaco, and also a second race in the same country, where previously only one per country was allowed, the first being at Watkins Glen, NY in the fall. FIA required that a sample race be held first; a F5000 race to held in 1975 and if it worked well, a Formula One race in 1976.

Chris Pook and the track plan

Forming the Long Beach Grand Prix Corporation, with race driver Dan Gurney set to design the circuit, Les Richter, formerly of Riverside Raceway and Ontario Motor

Speedway (Les later dropped out and was replaced by former world champion Phil Hill) and Long Beach attorney Don Dyer, they raised the $400,000 required by the city. The city committed to street improvements and the related costs for the event. They were able to form a semi public corporation with investors meeting certain standards; they had to be residents of California and have an annual income of $75,000 and were allowed to sell 50,000 shares of stock at $25 each. The Chamber of Commerce formed the "Committee of 300", volunteers who paid $175 to join and received jackets and ties. Pete Biro and Hak Ives, well known in the motorsports community came on board to hand public relations and Toyota was convinced to be the Official Car and Pace Car. There was some opposition; Deke Houlgate, the P.R. guy at Riverside Raceway, wrote an article in the L.A. Herald Examiner about how dangerous a street race could be. Fortunately it was countered by long time LA Times motorsports writer Shav Glick, who's L.A. Times article created a lot of interest overshadowing some of the bad press.

Long Beach Enters Grand Prix Picture

BY SHAV GLICK
Times Staff Writer

Signs on the approaches to Long Beach herald the beach community as The International City.

Long Beach brought in the Queen Mary to give it a touch of England. Now it proposes to become the Monte Carlo of the West by running a Grand Prix automobile race through downtown streets.

A U.S. Grand Prix West has been sanctioned by the Federation Internationale de L'Automobile (FIA) for April 1976. The Long Beach City Council voted unanimously to support the race, with full backing of the Chamber of Commerce, Convention and News Bureau, and the Downtown Long Beach Associates.

A shakedown race over the approximately 2-mile course will be held Sunday, Sept. 28, with Formula 5000 cars competing as part of the Sports Car Club of America national series.

The course, which has been approved by FIA representatives, starts on Ocean Blvd., near the intersection of Long Beach Blvd. It heads east to Linden Ave., takes a hard right turn toward Shoreline Dr., then sweeps east again along the beach in the shadows of the International Towers and Villa Riviera. At the intersection of Shoreline and Alamitos Ave., a hairpin turn heads back down the opposite side of Shoreline about a mile to Chestnut Pl. There, another hairpin turn brings the cars to Pine Ave., where a sharp left-hander runs past decaying shops and the old Pike and up a steep hill to Ocean Blvd. For another right-hand turn toward the start-finish line.

★

The interior of the course, most of which is on tidelands fill, includes the Long Beach Arena, Rainbow Lagoon and the under construction Convention Center and Auditori

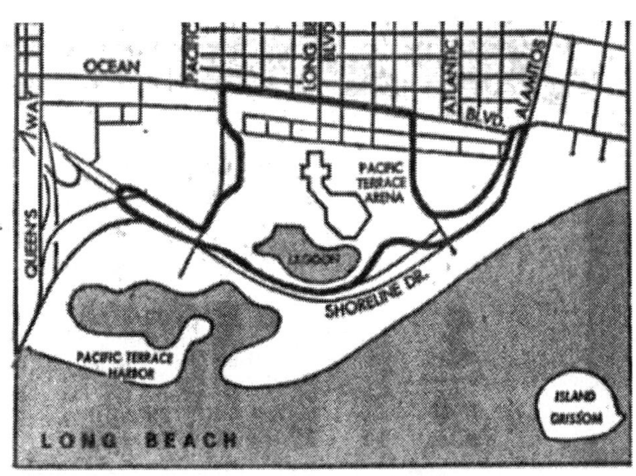

Proposed course of Long Beach Grand Prix West.

West. All that remains is getting the precise date on the FIA calendar. Long Beach Grand Prix Assn. officials will make a request for April 11, with April 4 or May 2 as alternate dates.

STOCK CARS—Jimmy Insolo, winner of last month's Permatex

In February, 1975, Indy 500 winner Bobby Unser was recruited to do a test drive in a Gurney Eagle Formula 5000 car on Shoreline Drive. He got arrested for speeding on the way from Long Beach Airport, arrived in handcuffs, the result of carrying an unregisted handgun. All was sorted out, Unser blasted down Shoreline but missed turning around and drove a ways up the 710 Freeway,

And so, on September 28, 1975, the first Long Beach Grand Prix, for F5000 SCCA / USAC cars was held, was a success and followed by fifty years of racing, the longest duration of any street race in the United States.

Read Chris Pook's Book To Learn More About Him. You'll Find It On Ebay and Other Places

The Improbable Messiah

How Christopher Pook became the savior of street racing and a city by the sea

By Leon Mandel

Mother Pook's Race Architect Pie
(Serves as many as the place will hold)

Two heaping measures Walter Mitty	*One pinch Rube Goldberg*
One lux Willie Sutton	*One tank car Kaiser Wilhelm*
Three tablespoons H. Ross Perot	*Essence of Edison, a mere whiff*
One quart Flo Ziegfield	*Generous measure Lewis Mumford*
One (brimming) Gallon Professor Harold Hill	*A little pinch of Ponzi*

Dump ingredients into baking pan. Stuff into oven. Leave for four or so decades. Peer inside very, very carefully. Do not singe eyebrows. Close door firmly. Call business rival. Tell him a real treat is waiting. Leave town.

It rises like a planner's idealized model from the wreckage of a toy city which has been randomly scattered by the hand of a malicious child.

Except the model and the city are real, and the new buildings are now so numerous that the outlines of the future can be seen growing from a Southern California landscape still dotted by great stucco edifices in pink and in white, their quatrefoil windows posing modestly behind decorative grilles, their roofs covered in tile and surrounded by parapets and topped with bell towers.

Long Beach used to be a sagging island waystop for the elderly, pierced here and there by volcanic tips of enlisted men's bars and porn movie houses that catered to the town's major industry, the Navy.

People lived in Long Beach, in houses like those described a quarter of a century ago by Raymond Chandler: "It was a dirty brown house with a dirty brown lawn in front of it. There was a large bare patch around a tough-looking palm tree. On the some intergalactic council. In fact, they are at their most dramatic when the event that pulled Long Beach into the present, the Grand Prix, comes to town.

Wally Edgerton, vice mayor of Long Beach, said last year. "The Grand Prix has been one of the great dynamics in the growth of Long Beach." Mario Andretti, elegantly tailored in a dinner suit for the Grand Prix ball, sitting one place away at the admiral's table at the San Pedro Officer's Club overheard. "The growth is more impressive than that. It is almost awesome to look around and see how the city has changed.

"In a way, it is a legacy of motorsports that all of us can be proud of."

This would be the same Mario Andretti who won a grand prix and two Indycar races through the streets of the city. That he was there to do so was thanks to the fact that Long Beach was perhaps the only town in America naive enough to believe in a transplanted Brit who emerged in 1974

And while you're at it, ask them how a man once described as "...variously thought of as...the inventor of urban racing, Rasputin of downtown speed ghettos, high-binding huckster who tap dances so fast you see neither the brass on his Ballys nor the hand picking your pocket..." came to be the president of the Chamber of Commerce of the city of Long Beach.

II

"There's a cadre of guys who look at me and say, 'Who is this guy?' (Some of them) think I sail pretty close to the wind."

Christopher Simon Robin Pook was born in England. That is established. Much of the rest of his early history is not. According to one account, a certain Miles P. Shook, citizen of Long Beach and apartment building owner, sponsored Pook's entry to the US at the request of his—Shook's—sister, a friend of the Pook family.

Pook himself talks about his business background as being in "construction: airplane charter, used equipment, and bro-

Long Beach Races for International Image

Hopes Grand Prix to Be Held There Will Boost Stature

BY JERRY RUHLOW
Times Staff Writer

Can a city that only two years ago banned electric cars from downtown sidewalks find fame and fortune as the site of an international auto race?

Long Beach thinks so, and is counting on the International Grand Prix next March to bring about a change in image the city has long sought through other means.

In recent years the beach city spent at least $60 million to buy the Queen Mary and turn it into a hotel and tourist attraction. It developed the most prosperous harbor on the West Coast. Millions of dollars are being pumped into downtown redevelopment. A $50 million auditorium and convention center are under construction.

But reputations die slowly and to much of the world, Long Beach is still "Iowa West," where Leisure World residents go to retire, where a Sunday concert in the park (by the nation's only full-time municipal band) is big-time entertainment.

Even adopting "The International City" as its official slogan hasn't done much to convince the world that Long Beach is a metropolis to be reckoned with.

What is needed, city fathers agree, is something dramatic, international in flavor, that will focus worldwide attention on Long Beach as an exciting place to spend a vacation or hold a convention (after all, San Diego has done it, even without the Queen Mary).

Enter Christopher Pook and company with the Long Beach Grand Prix, intended to turn Long Beach into the Monte Carlo of the West.

Ever hear of Emerson Fittipaldi of Brazil, Niki Lauda of Austria, Jody Scheckter of South Africa, Ronnie Peterson of Sweden, Jean-Pierre Jarier of France, Clay Regazzoni of Switzerland, Carlos Reutemann of Argentina or Lella Lombardi of Italy?

These are the all-stars of auto racing, and they have already signed up for the Long Beach race. Pook is confident that close to 100,000 racing

PROMOTER—Christopher Pook, with Shoreline Drive, part of Long Beach's *Grand Prix* course, and the Queen Mary in background.
Times photo by Jerry Ruhlow

buffs will turn out for the Long Beach Grand Prix in March.

Even the shakedown race—a Formula 5000 event scheduled to test the two-mile course through downtown Long Beach on Sept. 28—has attracted entries from such noted drivers as Mario Andretti, Al and Bobby Unser, Johnny Rutherford, Danny Ongais, Jackie Oliver, Gary Gabelich, Brian Redman and Warwick Brown, and is expected to draw an attendance of 75,000, Pook said.

Sanctioned by the Federation Internationale de L'Automobile (FIA), the Formula I race is one of 16 in the world. Others are held in Argentina, Brazil, South Africa, Belgium, Monte Carlo, Spain, France, Italy, Germany, Holland, England, Sweden, Austria, Canada and Watkins Glen, N.Y.

The Grand Prix-West will make the United States the only nation with two such races and it will be one of only two such races in the world run on city streets rather than a race track. The other is at Monte Carlo.

The course begins on Ocean Blvd., near the intersection of Long Beach Blvd., goes east to Linden Ave., takes a hard right turn toward Shoreline Drive, then goes northeast on Shoreline.

At Shoreline and Alamitos Ave., a hairpin turn heads back down the opposite side of Shoreline along the beach to another hairpin turn that will bring the cars to Pine Ave. A sharp left on Pine will take them up a hill to Ocean Blvd. for another right turn to the start-finish line.

Pook said it will be a relatively slow course, but cars will still hit 165 m.p.h. on the straightaway, with an expected average speed of 93 m.p.h.

Formula I cars are single-seat, open-cockpit and open-wheel vehicles with rear engines and a maximum engine size of 3 liters. They burn gasoline and have a five-speed gear box. Formula 5000 cars, racing in September, are bigger, but slower.

"The idea is to test the car and driver over a circuit encompassing all that would be met in street driving," Pook said.

"Drivers are truly remarkable and must memorize the course in advance. They're like computers. They can't see half of the turns until they're there, and then must program legs and arms to brake, downshift and make the turn while their mind is already working on the next turn."

But just how was the Long Beach Grand Prix born, and what will be its impact on the "International City?"

Pook, 34, who was born and educated in England but spent most of his youth in Spain where he raced as an amateur, directs the operation from a penthouse office atop the 600 Ocean Blvd. Building, commanding a view of the race course and the nearby Queen Mary.

A former tour operator, Pook said he conceived the idea two years ago after a luncheon with Bob Lichtenhan and Paul Ressler, managers of the Long Beach convention and news bureaus, respectively.

"They told me about the new convention center that is being built and we discussed the problem of bringing people to Long Beach, convincing them it was a fun place to go.

"Ten days later I came up with the idea for the Grand Prix. Auto racing, tennis and soccer are the only truly international sports, and Formula I racing cuts the swath through every income and educational bracket. It is recognized throughout the world and would present an opportunity to gain overnight international recognition for the city.

"First I contacted (racer) Dan Gurney, who was a little skeptical, but liked the idea and suggested we talk to (racing manager) Les Richter. Richter thought we were both crazy, but then became involved and the idea grew."

Other founders and board members are Long Beach attorney Don Dyer and stockbrokers John R. Queen Jr. and Stewart C. Elner.

The six principals have invested about $40,000 of their own funds and a lot of man-hours in getting the race organized, Pook said. Additional funding is being arranged through a special intrastate public offering to raise from $860,000 to $1,575,000 for operational and organizational expenses.

Please Turn to Page 2, Col. 1

Assorted Comments About the LBGP

(Note: If your comment didn't make it here, it's just because we ran out of room)

Anatoly Arutunoff: ***"I saw an ad, I think in Autoweek; an announcement that a group was forming to develop the idea of the Long Beach Grand prixbgp. They were looking for $750 (I'm pretty sure of that amount) to be considered a founder of the event. One way or the other I wound up in a phone conversation with a Mr. Queen, during which I told him that the USA racing colors were blue and white. He said they were considering sports jackets for the founders and he would suggest that they were blue. We kicked around the idea of wearing white slacks but decided that anywhere near cars, etc.,they'd get dirty. The great gold thread embroidered founder badges came later; I think we talked more than once. That first notification simply rang my bell; I was instantly certain that it'd be a success. The original plan was four seats for ten years, with two coming with extra privileges, such as meals, snacks, bar etc. I can't remember exactly how far it was into the decade that they decided to extend to the founders what they gave the high dollar investors: those benefits in perpetuity!! Best investment proportionally that I've ever made. Several years ago they started billing me for a ticket handling charge; a couple years ago I asked the nice lady who'd called me about this if this was a way of seeing if we were still alive and she said yes!!"***

Charley Budenz: "I went to the first race, never seen a street race before though I'd been to many racetracks; Riverside, Seats point, Portland, Seattle and raced on a few. What an excitubng event, formula 5000 was great though downtown Long Beach was pretty shabby. Wnt back often to Formula one and Indy Car; the transformation of Long Beach and the crowds the race brought in were a credit to Chros pook ans his determination, I was friends with Danny McKeever who trained all the celebrity drivers, he got us all passes to hanf with the rich and famous. Loonking forward to next year !!"

Pat Childs: "Long Beach has been one of my favorite races since I joined the Indy Car circus, in 1995, servicing tires for Bridgestone / Firestone. My first Long Beach event saw me covering Indy Cars, Indy Lights and the IMSA Bridgestone Potenza Supercar Championship, from our three service trucks, which were set up in three separate ***paddocks. Too many stories to recount but one stands out - In 2013, I was working behind pit lane and saw someone surrounded by Mexican media and television cameras, headed my direction. I had moved up against the fence, to let the mob pass, when I saw it was Adrian Fernandez, whom I had worked with back when Firestone first returned to IndyCar. He had just been inducted into the Long Beach Motorsports Walk of Fame. He spotted me and his face lit up as he made a bee line towards me, we shook hands, and I congratulated him then we made a bit of small talk as the cameras rolled. Apparently, I was famous in Mexico that day."***

Bobbie Cooper: ***"After the first race I asked my boss at DAECO, Harold Daigh, why we didn't provide the fuel. He just shrugged his shoulders so I asked if he would mind if I gave it a shot. He kind of laughed and said go ahead. I was a twenty something office***

clerk at the time, too young and naive to know this was something I couldn't do, I made an appointment with the VP of Operations. Our meeting was about an hour long and I walked out with a hand shake deal to be the official and exclusive fuel for the race.. Dealing with the F1 teams was always interesting. They all carried brief cases full of $100.00 bills. They would call me the fuel lady. I also drove in the first Toyota Celebrity Race, then they decided they wanted real celebrities so they retained me as their race driving instructor."

Sandy Dells: *"First time I ran the Atlantic race was 1982, the last year of F 1. I rented a motor home because of lack of hotels at the time. Right where the 405 and Long Beach Freeway came together a semi ran a car into the center divider in front of me, the car came back and kit me and I totalled the rented motor home, Race was a hoot with Willy T. on the pole. He crashed on first lap. which made it even better. I finished about 10th."*

John Dillon: *"I started flagging 35 years ago - Long Beach was already undergoing a massive transformation; major construction projects during those times forced several changes to the track layout over the years and the hazards of construction sites (rebar, loose earth, debris) wounded more than one of our volunteers. Flagging techniques were different too; back then you could stand at the edge of a wall to pop out flags, nearly bonking the drivers on their helmets. Often we would play in traffic to push stuck cars clear of the track or respond to fires. Now of course, none of that is allowed. As the event grew, race control grew even more. What started out as just a handful of people in a trailer with radios is now a room full of people and dozens of TV monitors and computers. One of my favorite stories is about the year Jay Leno raced in the Celebrity event. Put a dozen identical Toyota Corollas on the track at once, with a mix of amateur and pro drivers; what could go wrong? Jay was a spokesman for the FritoLay company, selling Doritos. During practice he crashed his Corolla, and crashed another during qualifying, so on the pace lap for the race, we showed him Burma Shave signs all around the track that together read crunch all you want Jay, we'll make more. The crunch all you want reference was the Doritos advertising tag line at the time. Afterwards he said he almost crashed on the pace lap he was laughing so hard.*

Peter Farrell: *"Long Beach was tight and unforgiving of course being a street circuit with concrete barriers and no room for error. But fairly smooth from memory. At the IMSA Supercar race in 1994, David Donahue won BMW Dinan M5 with. Shawn Hendricks second in a 300ZX Turbo and I was third in an RX-7. For me, the highlight of racing here was the massive crowd as it was a support event to CART. I remember chatting with Paul Newman on the dummy grid before the race while thousands of fans were screaming his name from a few feet away. Of course they had no interest in me. When the the green flag dropped Paul and I raced into turn one and I looked out my passenger window at his Lotus X180R right beside me. The thought in my mind at that moment was your fame and fortune won't help you now Paul as I late braked him into the turn and shut the door on him. I went on to fight for the lead and Paul finished eighth."*

Jim Gustafson: *"I was working at Jamestown Motor Center in Long Beach, hanging out at a racers hangout bar called the Tappet; another hanger on made me an offer on Lola T-190 F5000 car that was too good to pass up, later moved up to March 73A in time for the first LBGP. And it was epic - The roar of the engines echoing between the buildings on Ocean Boulevard; marching bands, crowds, Hollywood celebrities and more. Prior to practice, we got driven around the circuit in a Greyhound bus; Mario had the front seat !! Didn't do well, broke a driveshaft in turn one, the steep downhill right hander and that was my race. A great experience !!*

Stu Hayner: *"Long Beach was one of my favorites. My first ever Trans Am, I would have won when Les Lindley dropped out with just a couple of laps to go , but I was taken out two laps from the end by a backmarker. I finished fourth. Other times, the car that eventually won was sputtering coming out of all the turns. Not every race was without controversy, but that's racing."*

Lynn Hunting: *"My most fun times at Long Beach were as a corner worker. As press not so much, specially in the F1 days. My first year the infamous 'Long Beach Lunch' on Saturday got half the workers were sick overnight and Sunday."*

Mike Kojima: "Worked a display in the Expo with TRD, engineered for Falken and the US Air Force Teams in Formula Drift and Randy Pobst in Time Attack. Fun times but support people are boring."

Rick Knoop: *"Long Beach, one of my favorites. I've raced at a lot street courses, Columbus, Miami, San Antonio come to mind. Long Beach is the best. Plus I won the Historic Formula One race in my Tyrell, ran well in the Historic Can Am and hope here will ve a Historic GTP soon."*

Bob Lee: *"We had just finished assembling the Ti 22 a week prior to the Long Beach event. Tested at Willow Springs; headed to Long Beach; needed to modify the dry sump tank then unloaded the car and changed gear ratios. First session out we forgot to latch one door and Ilja Burkhoff, the driver, had to stop on the track and latch the door. He qualified fourth[h] behind three newer and fully dialed in cars so and we decided to make no changes to the setup because of time limitations. Ilja was getting out of turn 11 as well as the leaders but later admitted to lifting on the front straight because the car was drifting sideways. Turned out we were hitting the bump stops which explained the high speed stability problem. We finished fourth and were happy with that result."*

Joel Martin: *"I worked the first three races. At the F5000 race I worked as an observer on the land line. I was Assistant Clerk of the course at the first F1. Race control was in the basement of an old folks home who all had been moved to some place in Palm Springs for the race week. I monitored the land line and reported to Jerry Adams and Birdy Martin as Clerk of the Course. The TV people made a video tape of a couple on the balcony of the building across the street having sex. That was around for some time. The second year was pretty much the same arrangement. What I found there was the fact that we could work with the TV people in the room and have them roll back if there was something that needed more inspection. And one came up. A driver spun and stalled the car. The marshals went to move it to a safer place and as they pushed he dropped the clutch and took off. That was illegal and I had it rolled back and called Birdy. He agreed and told John Timanus who was on the radio in the pits to advise the car would not be scored.*

Next thing we know Colin Chapman and Eccelstone were in race control. Not the best time to meet them. The DQ stood. One other thing that stands out is the pass Mario made on Jody at the end of the long straight. Bob Swenson and I were on break and knew where to go to see the pass. We could have reached out and touched them at the apex."

Paul Marygold: *"There's a well known photo of Brian Redman, who won in 1975, where I'm in the background behind the tires. My job, holding the yellow flag watching down the hill from turn one to turn two. This turn, kinda' like the corkscrew at Laguna, fit the textbook definition of a blind corner. Watching the cars entering the turn holding the blue flag was Mike Pratt who also raced an Alfa Romeo in club races. There were more than a few times thar he grabbed and pushed me out of the way if it appeared a car would hit the tire barrier in front of us. I cite this turn as an example of the importance of the blue flagger having the yellow flaggers back. Mike and I were back in the same place for the first Formula One race in 1976. It was a few years later when James Hunt went airborne in turn one. Two of our flaggers, Maryanne and Pete, went in action. I say "our flaggers" meaning Cal Club regular flaggers, but the Long Beach race has routinely had flaggers from other SCCA regions and from other nations, Canada, the UK, Mexico, Netherlands, South Africa."*

Raphael Matos: *""Long Beach is a huge part of my racing career and I had phenomenal races around that place. Having a win back in 2007 in the Atlantic series was definitely the highlight. I always enjoyed Long Beach, the waterfront scene is epic and the track layout promotes incredible action. There's a reason why they called it Monaco of America"*

Mark Mitchell: *"i never got the chance to race there unfortunately but did the usual stuff in the early years. I remember climbing up the fire escape on an abandoned apartment building on Pine with a friend, got on the roof and could look straight down into the cockpit of Formula One cars and see drivers blasting up that short straight and then right on to Ocean past all the adult theatres One year we snuck in to the ventilation ducting that ran above the convention center inside the arena where the car were paddocked and could look right down on the action. Good times !!"*

Sam Moses: *"I was there for 17 or 18 years, probably 18 SI stories. Favorite memory is at the start, I think it was the first year, standing at the guardrail in pit lane, engines screaming, porn shops on the other side of the street, look up on a balcony and there's literally a long haired Jesus freak waving a sign about the End is Near, or something like that. I may have been was stoned, added to the surreality of the moment."*

John Norris: *In 1983 I got to work as crew for the Pro Celebrity race. We picked drivers out of a hat. I got Paul Moyer, Parnelli Jones and Dan Gurney. When I met Dan he walked up to me, shook my hand and said "Hello my name is Dan Gurney" I was thinking "Of course you're Dan Gurney everyone knows that! You are the only American to win a Formula 1 race in a car he built!" Anyway he went on to win the race at Long Beach. A couple of weeks later I got a letter from him thanking me, a bottle of booze and a check for $200! The check and the booze are long gone but I still have the letter. RIP Dan Gurney a truly great man. I'm so glad I got to meet you."*

Randy Pobst: *"I love the unusual, dangerous, exciting, nutty street courses, and always do well, if I don't totally destroy my car. Long Beach had unusually long straights, including the curving but flat out front straight, that lead to very high speeds. Also very unique is curving around the fountain with its veggie garden curbing. Always loved mowing the bushes with the splitter around the fountain. And the last corner hairpin, tightest of any I have ever experienced, except perhaps at St Petersburg on the pier. I really like the stadium feel of the last sweeper, lined with grandstands, and the smoothest part of the circuit every year. I outbraked myself going for the lead into the back straight runoff in '95 with BMW M5; still think I could've made it if I'd tried the corner. With those unrelenting walls and lost power steering in the K-PAX Volvo beastie, I would've been okay, except for that tightest ever hairpin. Just could not get another twist of the wheel, became a hazard, and had to retire.*

Brian Redman: *"I won the first year in Formula 5000,, surprised that we were racing in a slum with porn theaters and tattoo parlors. Invited back ten years later for the Toyota Celebrity Race; what a shock, no porn nearby, new condos and shiny buildings, a renaissance !!"*

Jim Sofronas: *"The Long Beach Grand Prix hold a special place in my heart for sure; our overall win in 2009 in World Challenge was a big boost for our business at GMG. The event is one of the best, not so much because of the track itself, it's more about the atmosphere and location. SoCal provides one of the best backdrops for all of racing and you can't beat the weather!"*

Lance Stewart: *"I ran the GTU race 1990 - Cool suit fails on the grid - Race starts and I got spun out by a GTO car passing early in the race. First couple laps I'm dead last. GTO cars are getting ready to put me a lap down. Caution comes out; restart happens with me the first car Because of the layout, cars are slowed way down approaching the front straight so I get the green and take off. Came all the way back to finish second behind my teammate, John Finger"*

Doug Stokes: *"I read about the race in Autoweek and, six months before the race, went to the Breakers Hotel and rented a room for the race weekend. The desk clerk seemed surprised when I requested a third or fourth floor room facing the street, rather than an ocean view. Worked on Bill Simpson's car; he broke everything, didn't get to race but my friends and I had a great place to watch from."*

Mark Vaughn: *"I used to live in San Pedro in the '70s, when the Long Beach Grand prix was still Formula One. My friends and I would ride our motorcycles over the Vincent Thomas Bridge and look for a place to see the race. Halfway through they stopped taking tickets and anyone could walk in! So we did. Forty years later or so I told the great Chris Pook that story and he said, "You owe me twenty bucks."*

Mitch Wright: *"I raced the Trans Am race in 1993, I don't remember where I finished but do recall being in awe of just being in the show, the crowd and the atmosphere, It was amazing and I thought I did well for an old guy."*

22
5
FLY
2

1975

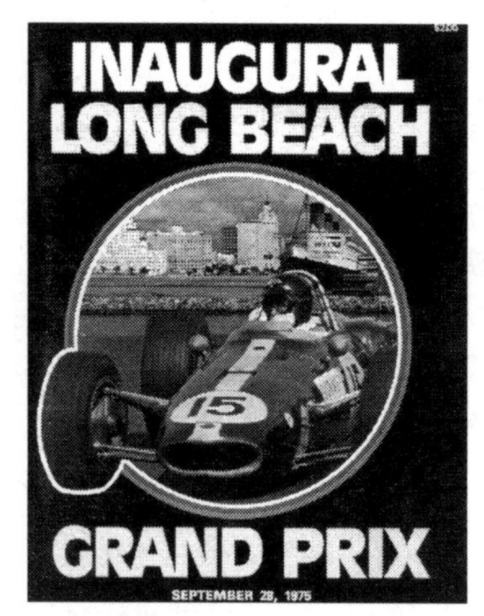

The 1975 Long Beach Grand Prix, race seven of the 1975 SCCA / USAC Formula 5000 Championship, was held on September 28, 1975 on a 2.02-mile (3.25 km) temporary street circuit in downtown Long Beach. It was the inaugural running, used as a test event for Formula One coming in 1976. The Formula 5000 cars turned out to be faster than Formula One cars !! World Champion Dan Gurney helped design the circuit; Phil Hill served as race director and Tom Binford as chief steward. Binford was the Indy 500 chief steward and president of ACCUS (Automobile Competition Committee For the United States). The announcer, Sandy Reed, was a fixture at Ascot and Riverside. A sidebar issue - over 95 people, local residents, who claimed to have been inconvenienced by the race were given a free trip to Catalina or San Diego. CBS televised the race a week later.

The Friday prior to the race, a kickoff luncheon was held at the Long Beach Arena, featuring drivers, race cars and the Toyota Celica Pace Car on display and CBS Sports commentator Ken Squier doing interviews. Tickets were $3.50.

Saturday saw 200 runners compete in a 5000 meter run along the race course and a 50 car Concours 'd Elegance on Pine Avenue.. A local attorney attempted to file a restraining order to stop the race but was denied.

Grand Prix halt denied

A temporary restraining order, which would have halted the Long Beach Grand Prix Formula 500 race this weekend, was denied in Superior Court Wednesday morning.

Judge Max Wisot denied an attempt by Long Beach lawyer Floyd King to have the race stopped because, King claimed, barricades put up for the event would deny him access to his apartment and law office in International Towers at 666 Ocean Blvd.

Attorneys for the respondents in the action quickly pointed out International Towers is not within the race area and that access to and from the building will be available at all times during the racing event. They said the south side of Ocean Boulevard and Atlantic Avenue will be open during the three-day period which begins Friday.

Capt. Al LaRue, traffic commander of the Long Beach Police Department, said that although traffic will be slow on Atlantic Friday, Saturday and Sunday, residents of the area, their visitors and people wanting to do business there will be admitted.

King, who explained he wasn't aware International Towers was outside the area being barricaded with concrete and heavy wire, said in his 10-page complaint submitted Tuesday that his son couldn't get to school, his wife to college classes nor his mother-in-law to church during the race.

He filed the action against Long Beach Grand Prix Association, the City of Long Beach, Grand Prix President Christopher R. Pook and five John Does.

Don Dyer, attorney for the Grand Prix Association, said he tried to tell King before the hearing that his building is outside the race area but that King wouldn't talk to him.

And on Sunday, Dan Gurney's father, John, an accomplished opera singer, sang the National Anthem. Then, a crowd of over 65,000 spectators first saw; at 2:00 PM, a match race between racing great and race director Phil Hill and racing champions Bob Bondurant, Dan Gurney and Graham Hill in identical Toyota Celicas. Phil Hill won.

Mario Andretti was the pole qualifier in the Parnelli Jones Lola T332, followed by Al Unser in the sister team car and Tony Brise in the Theodore Racing Lola T332. Qualifying times were set in two sessions on Saturday. Seeded drivers had one 45 minute session and one 1 hour, 15 minute session, while non seeded drivers had two 45 minute sessions. Starting positions were set by two heat races, with the finishers from heat one starting in odd positions and heat two in even positions. Tony Brise won the first heat and Al Unser the second.

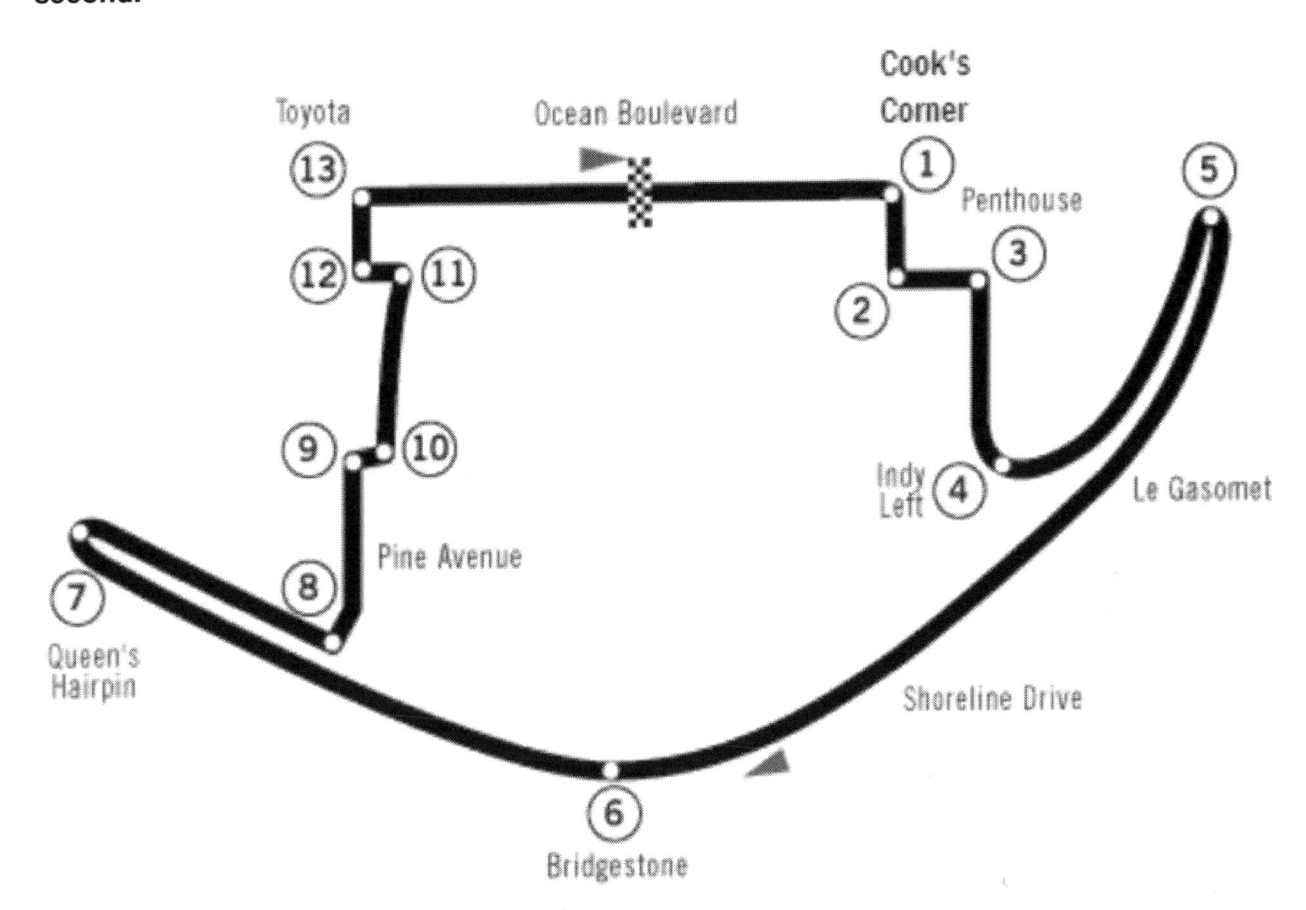

At the start of the race, 3:45 PM, Al Unser took the lead from second on the grid. On lap three, polesitter Tony Brise took over the lead with Mario Andretti in second. Brise led until lap 15, when both Andretti and Unser got by. Brise re passed Unser on the next lap, and battled with Andretti until he re took the lead on lap 29. On lap 33, Andretti retired with transmission problems, and one lap later Brise retired with a broken driveshaft.

Brian Redman, in his Boraxo T-332 Lola Chevrolet, won with a margin of 29.446 seconds over Vern Schuppan in the Dan Gurney prepared Jorgensen Eagle. Eppie Wietzes was third. Redman also clinched the season points championship for the second time. Redman earned $30,750 for the win. Redman was forced to adopt a conservative race strategy due to handling and differential problems and then capitalized on the misfortunes that befell early race leaders to claim a stunning victory. Redman said *"I don't like to win in this fashion. I'll be the first to admit that I backed into it. But winning is always nice, no matter how one attains it."*

(See all the race results are on the attached DVD)

65,000 see Redman win Grand Prix

RACE PHOTOS PAGES A-5, C-4

INDEPENDENT

WEATHER Mild weather today with some night and morning low clouds. Highs in mid 70s. Complete weather on Page C-11.

44 Pages LONG BEACH, CALIFORNIA, MONDAY, SEPTEMBER 29, 1975 Vol. 35, No. 39
HE 5-1161—Classified No. HE 2-5959 ●★ Home Delivered Daily and Sunday—$4.00 Per Month

Andretti, Unser fail to finish race

By ALLEN WOLFE
Staff Writer

British driver Brian Redman, forced to adopt a conservative race strategy due to handling and differential problems, capitalized on the misfortune that befell early race leaders Tony Brise, Al Unser and Mario Andretti to claim a stunning victory Sunday in the Long Beach Grand Prix.

It was the first street race in the United States in more than 50 years.

"I don't like to win in this fashion," said the proper Yorkshire gentleman. "I'll be the first to admit that I backed into it. But winning is always nice, no matter how one attains it."

Redman and his red, black and white Boraxo Lola T-332 Chevrolet won by a margin of 29.946 seconds over second-place Vern Schuppan of Australia, driving a Jorgensen Eagle prepared by Dan Gurney.

TAKING NOTHING away from Redman, his $30,000 march to the pay window was rather incidental to the scope and magnitude of the drama in which he emerged the triumphant hero.

A boisterous, festive, churning crowd—estimated by Long Beach police to be 65,000—turned out on a balmy autumn day to be a part of "Monaco West," as it had been described in month-long pre-race publicity.

They completely enveloped the 2.02-mile, 12-turn circuit that begins at the intersection of Long Beach and Ocean boulevards and winds in a snake-like loop down Linden Avenue, along Shoreline Drive, up Pine Avenue and back to Ocean. The grandstands lining the course were packed to capacity, as were balconies and windows along hotel row starting with the Breakers and running to the International Towers. Still others clung to roofs and theater marquees, one of them ipecuously featuring the X-rated "Sodom and Gomorrah."

(Turn to Page A-4, Col. 3)

★ ★ ★

Racing history made in L.B.

By LOU GODFREY

The local police loaded all the prostitutes in the area onto buses and shipped them to San Bernardino for the weekend. The porno theaters were told to cover the "Debbie Does Dallas" signs with banners saying "Welcome to the Long Beach Grand Prix".

Pre-race activities to begin on Friday

By RALPH HINMAN Jr.
Staff Writer

A series of major events highlighting this weekend's Long Beach Grand Prix Formula 5000 race begins here Friday with the public's first chance to see both drivers and cars during a kickoff luncheon in the Arena.

Hours later a group from the American Cancer Society's Long Beach chapter will open a two-day Grand Prix-related "Race Against Cancer" at the Pike. Available then will be more opportunities to get acquainted with the racers.

And on Saturday morning, Pine Avenue between 1st and 5th Streets is set to become the stage for a *Concours d'Elegance*, a Continental-style display of about 50 racing cars along with a third chance to meet drivers and photograph them with their vehicles.

A $1,000 cash prize awaits the driver, crew and car adjudged sharpest in appearance. Those buying $3.50 tickets to the 11:30 a.m.—1:30 p.m. kickoff luncheon may view cars on display and stay during the afternoon to watch trial runs over the full circuit. With streets in the area blocked for racing, entry to the Arena is via an overpass to be erected shortly before the luncheon across Ocean Boulevard near Elm Avenue.

In a different style seating arrangements, racing cars are to be driven through the Arena between 12:30 and 1 p.m. where they will be parked in front of the head table. Drivers then are to be interviewed by CBS sportscasters Ken Squire and Jane Chastain.

With the Committee of 300's Jim Willingham as master of ceremonies, head table guests include the Queen of the Grand Prix, Susan Matyja; Christopher Pook, Long Beach Grand Prix Association president; Dan Gurney, LBGPA vice president and veteran race driver; Phil Hill, rae director; Graham Hill, co-Grand Marshal; Dick King, U.S. Automobile Club executive director; Cameron Argetsenger, Sports Car Club of America executive director; Reynold McDonald, SCCA board chairman; Squire and Chastain.

Tickets remain on sale at the Chamber of Commerce, 50 Oceangate Plaza, and in many banks or other businesses here that display the Long Beach Grand Prix emblem.

The "other race" this week, a "Race Against Cancer," will be at the Pike Friday from 6 p.m. to midnight and between 4-12 p.m. Saturday. Sponsors are "Couples Against Cancer," a group organized here last to raise funds for the American Cancer Society.

Formula 5000 drivers and Grand Prix Queen Susan Matyja will meet the public between 7 and 10 p.m. Friday and again Saturday at 8 p.m. Drawings are set for 8 p.m. Friday and 10 p.m. Saturday in which the prizes are race pit and paddock passes, racing jackets and caps.

Tickets redeemable for a book of 9 Pike rides now are available for a $3 donation to the national society, its local chapter at 936 Pine Ave. Further information may be obtained by phoning 437-0791.

Racing cars will line Pine Avenue from 1st Street to 5th on Saturday between 9 a.m. and noon for a *Concours d'Elegance* Monaco style.

Drivers are expected to be standing by their cars from 11 a.m. to 12 noon in order to give autographs and be photographed with their admirers. And adding to a generally festive air during a morning of car viewing will be various combos from the Long Beach Municipal Band playing along the street. The entire band will concertize between 11 and noon.

And at 11, a $1,000 cash prize will be awarded the car, driver and crew judged to be sharpest by a committee of five downtowners — Wayne Christensen, Sam Rue, Jim Watters, Joyce Johnson and Roy Jarvis.

The award is to be presented by Rue, Downtown Long Beach Associates president. Participating with him in an event scheduled to be held at a bandstand near 320 Pine Ave. will be Grand Prix Queen Susan Matyja and Mayor Tom Clark.

During the morning Pine's two east lanes will be closed to traffic. Traffic will flow along the three west lanes except when the Formula 5000 cars are moved by carrier to their inspection positions.

The grand finale comes at noon, when drivers are set to start their cars preparatory to returning to the race area. Officials will help remove barricades and the cars will move south on Pine, west on 1st Street, south on Pacific Avenue, west on Ocean and again south on Pine to the paddock area.

Long Beach, Calif., Mon., Sept. 29, 1975 INDEPENDENT (AM) PRESS-TELEGRAM (PM)—C-3

Redman—'Nothing tops this'

(Continued from Page C-1)

mechanics, whom he said "did the greatest job of keeping cars together and ready for the main event than at any time that I can remember."

"We've never had a weekend like this in racing, and the attrition of

for the Formula I cars, though."

Redman then as much as said that he would not return to Long Beach for the March 28 Formula I event.

"I'm not interested in Formula I cars," he claimed. "I began my dislike for those cars in 1968

They went over it time after time, and even when everything seemed okay, they went over it again."

Like most Grand Prix drivers, Redman had the highest of praise for Long Beach's first street race.

"This event was absolutely incredible," bubbled the champion. "The gener-

houses all along the race course. I guess those are for the old people they say flock to Long Beach.

"I have seen a lot of old people here, but I think this race brightened their lives."

Asked if anything topped his Long Beach Grand Prix win, Redman

reason to quit racing now. But when racing is no longer fun, I hope I have the moral courage to stop."

The gracious Redman, considered not only one of sports car racing's all-time greats but also a gentleman both on and off the

L.B. Grand Prix thrills throng

BRIAN REDMAN SPRAYS CHAMPAGNE OVER CREWMEN AFTER GRAND PRIX VICTORY IN L.B.

Brian Redman zooms to win

By ALLEN WOLFE
Staff Writer

British driver Brian Redman, forced to adopt a conservative race strategy due to handling and differential problems, capitalized on the misfortune that befell early race leaders Tony Brise, Al Unser and Mario Andretti to claim a stunning victory Sunday in the Long Beach Grand Prix.

"I don't like to win in this fashion," said the proper Yorkshire gent. "I'll be the first to admit that I backed into it. But winning is always nice, no matter how one attains it."

Redman and his red, black and white Boraxo Lola T-332 Chevrolet won by a margin of 29.946 seconds over second place Vern Schuppan of Australia, driving a Jorgensen Eagle prepared by Dan Gurney.

Taking nothing away from Redman, his $30,000 march to the pay window was rather incidental to the scope and magnitude of the drama in which he emerged the triumphant hero.

A boistrous, festive, churning crowd—estimated by Long Beach Police to be 65,000—turned out on a balmy autumn day to be a part of "Monaco West," as it had been described in month-long, pre-race publicity.

They completely enveloped the 2.02-mile, 12-turn circuit that began at the intersection of Long Beach and

TOP CONTENDERS Tony Brise (64, in front), Al Unser (51), Mario Andretti (5, behind Unser) and Brian Redman (1, left of Andretti) hang together in early lap of Long Beach Grand Prix. Redman capitalized on bad luck of other three and emerged as winner of the 101-mile race.

—Staff Photo by TOM SHAW

Brown still smiling—and walking

★ ★ ★ ★ ★ ★ ★ ★ ★

Aussie pilot cheats death

By ALLEN WOLFE
Staff Writer

The scar tissue on the left leg runs in a haphazard maze from behind the calf and migrates to an ankle that appears as if it was pieced together by committee.

The scalpel incision on the right leg starts from just below the kneecap and parallels the tibia (shin) bone, ending in a crater-like hole above the ankle. The entire limb is bowed out, reminding one of the walking gait of a Texas ramrodder.

"I'll never win a lovely legs contest, that's for sure," says Warwick Brown with a laugh.

A bronzed 25-year-old bachelor from the unlikely hamlet of Wahroonga, Australia, Warwick can afford to be jocular these days. Instead of reclining in a wheelchair and staring at an empty left pant-leg, he's doing what comes naturally, driving a 550-horsepower Formula 5,000 car at speeds of 170 mph.

On Sunday he will climb unassisted into the cockpit of his red and white Norris Industries Talon MR-1 to compete in the $102,000 Long Beach Grand Prix. He has already won the greatest race of all—the race against death.

"I remember *that* day as if it was yesterday," he recalls. "February 4, 1973."

ON THAT DAY 32 months ago, Warwick was a competitor in the Tasman Cup series, the Australian version of the Formula 5,000 series in the United States. Like the other 20-odd drivers in the series, he entered the sixth round at Surfer's Paradise in the northeast province of Queensland. "It's the Miami Beach of Australia," he says. "Lots of girls in bikinis."

Midway through the race, the left front tire on his Lola T-330 deflated on a high speed corner and the car careened head-on into an earthen bank at 145 mph. The impact was so violent it pancaked the front nosecone and crushed the bulkheads supporting the driver's cockpit. The car somersaulted five times before coming to rest more than 100 yards from where it left the course.

It took course marshals and rescue workers 20 minutes to cut away metal fragments before they could extricate him.

His injuries were massive. The left leg was broken in five places and his crushed ankle was imbedded with slivers of metal and oil. The right leg was fractured in three places and he had a hairline fracture of a lumbar vertebrae.

"That night the doctors explained the situation to my mother," says Warwick grimly. "Under the circumstances, they thought the left leg was beyond repair and they wanted to amputate. But she knew how much racing meant to me and she refused to give them permission.

"Sometimes I think doctors seek out an easy solution. All my mother said to them was 'try to save it.' I'll always be grateful to her for the courage she showed under stress. I don't know if I could have done it."

Dr. Barry Bracken, resident in orthopedics, a man he describes as "the most dedicated doctor I've ever run across. If it wasn't for him, I'd be in a wheelchair today, or at best walking with a cane. I literally owe him my life."

Soon after Warwick was admitted, Dr. Bracken became concerned over the fact that his patient's left leg—the most physically damaged—was healing and progressing nicely. But the right leg was not keeping pace by generating new bone growth.

"The right leg was shorter and there was a definite possibility it would remain that way," says Warwick. "I couldn't bear the thought of being a cripple the rest of my life."

But there was one avenue of hope. The medical profession has known for years that bone growth is stimulated by electronic impulses from the cortex of the brain. Since the early 1960s researchers

tiny low-voltage, battery-operated unit directly into the large thigh muscle of Warwick's right leg. It was monitored closely for a period of six weeks, then surgically removed.

"The doctors were elated," he says. "They told me they had obtained the equivalent of six months natural bone growth in one-fourth the time."

Almost three months from the day of the accident, a gaunt-looking 5-8, 150-pounder was released from the hospital. But he still faced an uphill climb to rebuild strength in his shattered limbs, agonizing days spent swimming, walking, riding his motorcycle and falling off his surf board—all pointing to the day he could return to racing.

That day came in November, 1973, nine months after his near fatal mishap. "It was a non-competitive car and I ran only 10 laps," he recalls, "but it gave me the incentive to want more. I knew

L.B. Grand Prix goal: 'safest course in world'

JOHN DIXON, Sports Editor
TUESDAY, SEPTEMBER 9, 1975
SECTION C—PAGE C-1

By ALLEN WOLFE
Staff Writer

Dr. Peter Talbot was one of many course marshals stationed at critical points around Laguna Seca Raceway on the Monterey peninsula one day 10 years ago.

Suddenly a sports car went out of control directly in front of him, did several loop-de-loops and came to rest against the inner barrier of the track. The young driver vaulted from his crippled machine,

overlooking Shoreline Drive, surrounded by a sea of charts, maps, formulas, tables—and empty sandwich wrappers.

His title is safety director of the Long Beach Grand Prix Formula 5,000 race, a 50-lap, 110-mile dash through the downtown streets of Long Beach on Sept. 28. On his 59-year-old shoulders wrests sole responsibility of coordinating and directing all safety-related aspects of the $125,000 event.

It's a monumental task. Under his jurisdiction

Talbot is recognized as one of the world's foremost authorities in the field of race track safety systems and their application.

As late as five months ago he was "a consultant in an advisory capacity." Now the whole ball game is his.

'New' Shadow

Jackie Oliver pilots Formula 5000 car around Road Atlanta in first test of new Dodge power plant. Oliver's Shadow Racing Team will be only two cars not using Chevrolet power plants in Long Beach Grand Prix Sunday.

Dodge invades Formula 5000

By ALLEN WOLFE
Staff Writer

Hey, fellas, there's a new kid on the block.

In this case, the "new kid" is an engine, a 305-cubic inch Dodge developed by Chrysler Corp. engineers for the UOP Shadow team competing in the Long Beach Grand Prix Sunday.

Two of the 550-horsepower stock block V-8s have been installed in the Shadow DN-6 cars assigned to Jackie Oliver and Formula 1 newcomer Tom Pryce.

What makes the engines unique is the fact that they have only been tested competitively in one previous race —four weeks ago at Road Atlanta in the Wrangler 5000.

The powerplants are also bucking Formula 5000 tradition. Of the 49 cars

A 28-page special section on the Long Beach Grand Prix Formula 5000 race is featured in today's editions. The special section includes the official entry field, feature stories, pictures, map of the course, driver biographies and other informative data concerning the $102,500 race through the city streets Sunday. It's another bonus for I.P-T readers.

entered in the $102,500 race through the downtown streets of Long Beach, 47 are powered by Chevrolet engines produced by Los Angeles-based firms like Ryan Falconer, Bartz and Traco.

THE NEW Dodges are the result engine has provided the "go" for a Formula 5000 car.

The engine, slightly modified by the UOP Shadow team, incorporates many Chrysler high performance parts. Others were developed by the team.

"The engine showed considerable potential even in dynamometer testing," says Shadow team manager Mike Hillman. "Then we conducted a series of exhaustive tests on a small track to assure that it would be reliable."

Jackie Oliver further enhanced the team's faith in their product by finishing fourth at Road Atlanta.

"That was the ultimate test because the stresses and strains produced in racing simply can't be simulated," explains Hillman. "Our fourth place demonstrates just how competitive the engine is."

Hillman reveals that incorporating the new engine into the Shadow DN-6 did not present any major logistics problems.

"Only a few minor modifications were needed to fit it, such as accommodating the plumbing for its cooling system."

ANOTHER modification has been in the ignition system. In earlier versions of the DN-6, a magneto was used. Now a transistorized coil system has been installed. It caused some minor problems at Road Atlanta, but these were quickly corrected at track side.

There are no outward physical

Andretti has tips for drivers

Mario Andretti's advice to drivers in Sunday's Long Beach Grand Prix is to "be tidy."

Andretti should know ize that there are no fields to slide into harmlessly, and no places to nip off a little piece of track here and there to gain a little

But does he like the idea of "doing his own thing" right downtown?

"Like it? I love it. This kind of racing is the Real

Scheckter, Tom Pryce, Tony Brise and Jackie Oliver, have been to Monte Carlo and thus figure to have an edge over

PALMER TOYOTA

Proudly Presents the

GRAND PRIX PACE CAR

See it Now!!

ON DISPLAY ON OUR SHOWROOM FLOOR

OFFICIAL PACE CAR
LONG BEACH GRAND PRIX

Now you can buy one of these Special Edition "Long Beach Grand Prix" Toyota Celica GT models, loaded with special custom equipment. Hurry! Limited production.

PALMER TOYOTA

4401 E. PACIFIC COAST HWY., LONG BEACH

"On the Traffic Circle" 597-3686

See 50 Grand Prix Racing Cars

in Downtown LONG BEACH

SATURDAY SEPT. 27 FROM 9 A.M. to 12 NOON

IT'S ALL Free

ON PINE AVE.

The Grand Prix drivers and crew will be on hand with their cars. See them up close. A once-in-a-lifetime opportunity.

THE BIG NAMES IN RACING ARE HERE

• Andretti • Redman
• Unser • Oliver
and a host of other

Bring Your CAMERA

to photograph fabulous racing machines and their internationally famous drivers and crews.

Plus . . .

- ★ Autograph Session 11-12 Noon
- ★ Musical Groups
- ★ Long Beach Municipal Band

Downtown Long Beach is where it's at!

$1,000 Prize

plus 1st, 2nd and 3rd place trophies will be presented for best racing car display at the Bandstand at 320 Pine Ave. at 11 A.M.

LBGP

View the video on our you tube channel www.youtube.com/@racinghistoryproject2607 and read the Autoweek and Motor Trend articles on the attached DVD

Paul Marygold, working as a flagger, can be seen behind the tires as Brian Redman goes by !!

Mario Andretti

Bob Earl

A 'happening' hits Long Beach

Grand Prix practice wasn't perfect

By ALLEN WOLFE
Staff Writer

Monaco came to Long Beach Friday and the over-all effect was staggering—in more ways than one.

From a logistics standpoint, the coordination in staging the opening day of practice for the Long Beach Beach Grand Prix progressed as smoothly as a Marx brothers comedy. But, as Dan Gurney, chief of operations, pointed out early in the morning, "Considering the magnitude of this venture, the potential for problems is quite high."

He was true to his word. During a six hour period, starting at 6 a.m. the following occurred:

• Construction to place tire piles and reconstruct debris fencing along Alamitos Ave. was delayed until 8 a.m., forcing the opening of practice from 10 a.m. to 1 p.m.

• The delay, in turn, caused impatient spectators to gather in potentially unsafe areas and required 20 additional minutes for Long Beach Police Dept. officers to clear the troubled area.

• In the pre-dawn hours, a forklift hit an abutment supporting the Elm St. spectator overpass (bridge), temporary rendering it as "unsafe." The situation existed only two hours and then it was reopened.

• Spectators began filing into grandstands on Ocean Blvd. at the same time work crews were still tightening the supporting foundations.

• Two sets of work crews accumulated 32 consecutive hours on the course, erecting unfinished debris fencing, finally

SATURDAY Sports
'Best Sports Section'

JOHN DIXON
Sports Editor
SATURDAY, SEPTEMBER 27, 1975
SECTION C, PAGE C-1

The responsibility for the catastrophic series of events, quite naturally, fell on the 34-year-old shoulders of race president Chris Pook, who admitted that "now ment by the paste 'these are times that try men's souls.'"

But it is a testament to his cool reserve under stress that practice got under way at all, considering the monumental foulups. Pook said the biggest problem was the early morning construction-traffic flow situation at the intersection of Alamitos and Ocean.

"We had to keep this area open to people who work in the downtown area," said Pook. "It's a main artery and it's also a critical turn in the course. Once the traffic diminished, we faced the problem of getting it ready for racing. After all, the safety code calls for placement of 160 tire piles (stacked on moveable pallets)."

City manager John Mansell was almost immediately rescinded one of the major contract requirements between the city and the Long Beach Grand Prix Association, the proviso that the course be opened to public traffic between all practice and qualifying sessions. That means the course is now self-contained for the remainder of the event.

"He realized we were struggling," said Pook. "He told us to go ahead and make all the necessary adjustments to conduct the event safely and properly. Not having to remove the barriers the rest of the weekend is a big load off our shoulders and the sanctioning bodies (United States Auto Club and the Sports Car Club of America)."

All the pent-up anxiety and frustration of the day seemed to melt away

Benny Scott

Tony Brise

Warwick Brown

Skeeter McKittrick

John Morton

Jon Woodner

1976

The second Formula One race held in California, billed as the United States Grand Prix West, the first being at Riverside in 1960, was held on March 28, 1976. A purse of $265,000 and an additional $245,000 for travel expense made this an astounding $510,000 expenditure for the event. The city was to be reimbursed $281,000 for services provided. The Long Beach Grand Prix Association had raised over $900,000 through shares of stock. Sunday attendance was claimed to be 97,000. CBS televised the race a week later.

On the Tuesday night before the race, the Lions Club promoted a radio controlled miniature car race at the Long Beach Arena. Thursday was the Concours 'd Elegance on Pine Avenue and then on Friday the kickoff luncheon was held at the Long Beach Arena. On Saturday, a 10,000 meter foot race, an Olympic trials bicycle race and a historic formula one exhibition race filled out the day. The historic race featured drivers like Juan Fangio, Stirling Moss, Maurice Trintignant, Peter DePaolo, Denis Hulme, innes Ireland, John Surtees, Rene Dreyfus, Dan Gurney and Phil Hill, many driving rarely seen racecars. Gurney won in a 1959 BRM. Sunday, race day, featured a ten lap Kawasaki 750cc motorcycle race won by Yvon Duhamel and the Toyota Celebrity Race, won by Car & Driver's Don Sherman followed by the main event at 1:15PM.

KFOX radio, 1280 AM, covered the race with its Grand Prix Report on Wednesday, Thursday and Friday, interviewing drivers and celebrities and had race day coverage updates every fifteen minutes from journalist Max Muhleman and driver / builder Carroll Shelby

Grand Prix means money

L.B. race March 28 a $510,000 event

By ALLEN WOLFE
Staff Writer

Automobile racing is big business and nowhere is it more evident than Formula One Grand Prix racing.

Chris Pook knows only too well.

Early this week, the president of the Long Beach Grand Prix Association announced all provisions as stipulated in contract negotiations between his organization and the Formula One Constructors Association have been fulfilled, thus clearing the path for the inaugural United States Grand Prix West through the streets of Long Beach on March 28.

month (March 7)," explains Pook. "The extra money ($245,000) may seem prohibitive, but not when you consider the logistics and personnel involved."

The monies will be used to hire three British Caledonia Airways charter jets one week prior to the race, one carrying about 225 members of the Grand Prix circuit (drivers, mechanics, car owners,

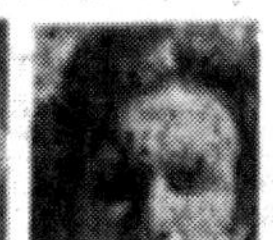
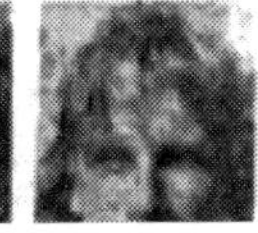

the Formula One Constructors must provide their own surface transportation.

Actually, the contract between the Long Beach Grand Prix Assocìaton and FICA isn't really a contract in the purest sense of the word.

"Bernie (Bernard Ecclestone) and his group are very informal in their dealings," said Pook. "We both have a letter and agreement of exchange, whereby they agree to perform and we agree to pay.

"I talked with Bernie on a number of occassions and we had a long discussion

This third round of the Formula One season was won by pole sitter Clay Regazzoni in a Ferrari with a 42 second lead over teammate Niki Lauda. Patrick Depailler was third in a Tyrell. Regazzoni earned $36,135=6 for the win. The Formula One Constructors' Association had decided to limit the field to twenty starters for safety reasons, because of the narrow concrete canyons necessitated by the street layout, and seven cars failed to qualify.

REGAZZONI WINS POLE POSITION

Continued from First Page

the Vel's Parnelli crew changed engines for the afternoon time trials. It was to no avail, however, as a broken left rear up-right bolt sent Andretti to the sidelines again.

Saturday's practice, and a Historic Antique Car race won by Dan Gurney in a 1959 BRM, attracted a crowd estimated at 40,000 by Long Beach police.

This is the first of two world championship races in the U.S., the first time since the series started in 1950 that one nation has been granted two races. The second will be the U.S. Grand Prix East, Oct. 10, in Watkins Glen, N.Y.

The Long Beach course, comparable to the famed Monte Carlo circuit in Monaco for its ocean-front venue and its relatively slow speed on narrow city streets, is the same as used last September in a Formula 5000 shake-

the course until the last 10 minutes of the one-hour sion. In the last frantic laps around the circuit he aged to improve his speed from 86.69 to 87.31, but m by two-tenths of a second recapturing the No. 1 po from Regazzoni.

Indicative of the close competition is the spacing o seventh, eighth and ninth qualifiers, Jean-Pierre Jari France, Vittorio Brambilla of Italy and John Wats Ireland in the Roger Penske-prepared Penske. Each speeds of 86.40 m.p.h. and officials had to go to the decimal point to determine their order. Jarier's Sha which circled the course at 1:24.163, was 7/1000ths q er than Watson's 1:24.170.

Today's purse is $265,000 and Grand Prix Assn. off need receipts of more than $1 million to break

At the start, polesitter Regazzoni led, ahead of Hunt. Vittorio Brambilla squeezed Carlos Reutemann into the wall, putting both cars out. Gunnar Nilsson's Lotus broke its rear suspension and jerked hard into the wall and caught fire. On lap four, Hunt and Depailler, battling for second, place, came together. Hunt yanked himself from his car, certain that it was undrivable, and shook his fist at Depailler each time the Frenchman came around. After the race, the McLaren mechanics came to retrieve the car and were able to drive it back to the pits!!

This would be the last race for the Vel's Parnelli car. Over three seasons, it competed in 16 races, with Mario the car's only driver. Upon retiring from the race in Long Beach, Andretti was approached by a television reporter in the pits, asking, *"How about this being your last race in Formula One?"* Andretti replied, *"What are you talking about?"* The reporter said, *"That's what Vel Miletich told me."* Andretti said, *"It may have been his last Grand Prix, but it won't be mine."*

Andretti terminated his relationship with Miletich and Parnelli Jones that day, but the next morning, by accident, joined Lotus team manager Colin Chapman for breakfast in a Long Beach coffee shop, where the two forged an agreement. By the next season, with Andretti driving Chapman's revolutionary Lotus 78, the two were winning races together and, of course, in 1978, captured the World Championship.

The first USGP West was a success. Indeed, former team manager Rob Walker said, *"I think the creation of the Long Beach GP was the greatest achievement in motor racing this decade".*

Drivers differed in their opinions of the concrete lined street circuit which featured two hairpins and a long, curving waterfront "straight." Ferrari's reigning World Champion Niki Lauda said the course was much bumpier than Monaco and harder on the car, but easier on the driver. Emerson Fittipaldi said he liked it very much, but Frenchmen Jacques Laffite and Patrick Depailler would not agree. David Hobbs said, *"The course is a bloody lot of fun to drive; it has a great deal of variety and you never get bored out there. But it's so damnably slow."*

Wire-to-wire in Grand Prix

Ferrari wins with Regazzoni; Lauda, No. 1, finishes second

LONG BEACH, Calif. (UPI) — Clay Regazzoni, the second banana to world driving champion Niki Lauda of Austria on the vaunted Ferrari factory team, said everything went according to plan.

In other words, he had only one thought in mind when he showed up Sunday for the first United States Grand

start," Lauda explained, "because I was just trying to take care of my car. I did try to close the gap later but I discovered that was too hard on my car.

"At the finish I slowed because I didn't want to wear out my brakes."

For Ferrari, it was victory No. 3 in

deliberately running him into the wall on the fourth lap. The Englishman, a winner of the noncounting Formula One race at Brands Hatch, Eng., (two weeks ago, had to leave the race.

"I think Hunt should race with people who are really racing," Depailler said

Sunday's event around Long Beach's 2.02-mile, 12-turn course, he was "very confident" he could win the race when the day started.

"I got a good start," he said. "After that, it was just a matter of finishing. No, I wasn't surprised. I felt very com-

(See the complete race results on the attached DVD)

City receives interim report on Grand Prix

By DON BRACKENBURY
Staff Writer

An audited interim financial statement of the Long Beach Grand Prix Association has been delivered to the city, as required under its contract, but is awaiting review by the Corporations Commissioner of California before being made public, City Manager John R. Mansell said Monday.

Mansell also said that the $180,005.44 which the association was billed for special city services, such as police and fire, for the Formula 5000 Race last Sept. 28 is not due under the contract until Feb. 1.

IN A LETTER which is on today's City Council agenda, City Auditor Murray T. Courson called attention to the fact that the city has not been reimbursed for the $180,005.44 in "extraordinary expenses" imvolving the Sept. 28 race,

Mansell agreed, but said the contract requires payment within 30 days of the city billing the as-

pay the $180,005 in a lump sum, or in several payments prior to the upcoming Formula I race on March 28 has not been determined, Mansell said.

The financial statement of the Long Beach Grand Prix Association, under the contract with the city, was due 120 days prior to the March 28 event, but Mansell said association officials said they could not meet that deadline because they had not received the city's bill for services. As a result, he said, his office granted the association a 30-day extension.

Jim Michaelian, controller of the Grand Prix Association, said the financial statement is as of Oct. 31, 1975, reflecting the position following the Formula 5000 race.

"BECAUSE WE ARE currently engaged in a stock subscription offering," Michaelian wrote, "it is the desire of the Corporations Commission of California to review this statement before its public dissemination."

Vittorio Brambilla

James Hunt

Patrick Depailler

Chris Amon

Ronnie Petersen

Clay Regazzoni

Jody Scheckter

Gunnar Nilsson

Tom Pryce

Jochen Mass

Sports Illustrated

The US Grand Prix West

Autosport

Long Beach Grand Prix

Japan's Le Mans car

Watch the video on our you tube channel - www.youtube.com/@racinghistoryproject2607
Read Motor Trend, Sports Illustrated, Autosport and Motor Sport on the attached DVD

The Long Beach Grand Prix Association announced that the race would be on in 1977 as they settled the $868,000 debt accumulated in 1975 and 1976 for .35 in cash and .20 in stock.

Long Beach plans Grand Prix again

LONG BEACH, Calif. (AP) — Despite financial woes in its first running reportedly accumulated for the first Formula I race in Long Beach and a

Hunt with Miss U.S. Grand Prix West

1977

The 1977 Long Beach Grand Prix, held on April 3, 1977, was the fourth race of the 1977 Formula One World Championship.

Race week kicked off on Thursday with the Prix View Luncheon at the Long Beach Arena and the Justice Brothers sponsored Concours d'Elegance on Pine Avenue. Friday was devoted to practice and qualifying by the Formula One cars but also practice for the motorcycle race and Toyota Celebrity races.

Saturday included more practice and qualifying but also a 10,000 meter run, a motorcycle race at 3:30 PM and fireworks at 8:00 PM. Sunday, race day, had a 50 kilometer bicycle race in the morning, the Toyota Celebrity Race at noon and the main event at 1:00 PM.

Sam Posey won the celebrity race and the professional class, followed by amateur Shelly Novak in second and Danny Ongais third..

70,000 watched as Mario Andretti, won driving a Lotus Ford, the only American to win a Formula One race on home soil, while also giving the ground effect Lotus 78 its first win. Pole qualifier Niki Lauda was second in the Ferrari and Jody Scheckter finished third in the Wolf Ford. Mario collected #55,670 for the win.

Niki Lauda on pole for Grand Prix today

The Daily Breeze — Sports — Sun, April 3, 1977 E1

Baseball previews on pages E8-9

By Paul McLean

For all Niki Lauda knew when he finished his final lap of the day, Mario Andretti had the day's best time, Jody Scheckter was second and he was somewhere near the top.

Now he knows better.

Lauda waited until his [illegible] on last lap Saturday to record the field's best qualifying time and win the right to sit in the pole position in today's Long Beach Grand Prix.

"It's a very difficult course to predict," said Lauda, winner of last [illegible] Grand Prix.

"The car is always jumping up and down. You run a quick lap and you do not know it. Then when yu think you're quick, it turns out you're slow.

"I didn't know I had the pole until I pulled into the pit and they haded me the sheet."

Lauda's final trip around the track was clocked at 1:21.[illegible] with an average speed of [illegible] miles per hour.

The average speed was better by about a second-and-a-half than last year's [illegible] Regazzoni. Most of the field, in fact, bettered last year's low qualifier.

Regazzoni averaged [illegible] Saturday, [illegible] miles per hour faster than last year's [illegible]. This year, he will start the race behind [illegible] other drivers.

Andretti, who had the low qualifying time for most of the [illegible], will start today's race next to Lauda in the No. 2 position.

Andretti led Friday's qualifiers with a time of 1:22.[illegible], [illegible] miles per hour, marks which held up until Jody Scheckter [illegible] miles per hour.

The time didn't last long, however, as Andretti followed it a few laps later with a 1:21.[illegible], [illegible] miles per hour.

Andretti said, prior to the qualifying, that the times would be improving.

"I think, like everyone else along pit row, we'll be scratching," he said. "The battle is [illegible]. There's more rubber on the track so there is considerably more traction. [illegible] be getting faster."

Several of the drivers lowered their times more [illegible] his Friday mark five times before winning the pole.

Andretti, however, couldn't keep it up.

"I couldn't quite squeeze any more out of it," said Andretti, whose best finish in Grand Prix races this year was a fifth in the Argentine Grand Prix, which was won by Scheckter.

Today, he said, the strategy will be to get into the lead quickly.

"If you're leading the race, you'll have to keep the lead," Andretti said. "If [illegible] to have to make your move as early as possible.

"I think it's going to be a pretty hard fought race. Everyone is extremely aggressive. All the factors seem to be there for an extremely close race."

Scheckter, who finished second to Lauda in the South African Grand Prix but is the leader in the Grand Prix standings, agreed with Andretti about the outcome of te race.

"It's so close," he said. "Lauda's a quick driver and he's got a quick car.

The Formula One crowd had arrived in mourning following the deaths of Tom Pryce at the South African Grand Prix and Carlos Pace in a light aircraft crash near São Paulo. The Shadow team signed Alan Jones as Pryce's replacement, while Pace's place at Brabham was taken by Hans Joachim Stuck. Stuck had been expected to drive for the new ATS team with its old Penske car but changed his mind, meaning that ATS had to sign Jean Pierre Jarier instead.

At the start, Scheckter shot past both Lauda and Andretti, and led into the first turn. Reutemann pulled next to Andretti on the inside approach to the turn, but braked too late

and slid straight on. Andretti avoided a T-bone by braking in time to duck behind him into the corner.

James Hunt was allegedly pushed from behind and when he hit John Watson's right rear wheel with his left front, he was launched six feet in the air, showing Watson the entire underside of his McLaren. On landing, he slid past Reutemann and down the escape road. Hunt was able to make it back to the pits, and though his suspension was bent, he carried on, and ended up missing a point for sixth by just two seconds.

"I was preparing for a real banzai under braking,"* Andretti later said. *"I needed to go from fifth to first gear in order to do it, and the way the gearboxes were in those days, I had maybe one or two tries to do that. My objective was to do it if I was in a position at the end of the race. Then I saw a twitch and, obviously, he had a tire that was slightly deflating."

On lap 77, Andretti outbraked the Wolf and pulled inside him entering the hairpin. *"It's not that he went wide,"* he said afterwards, *"I just got him clear at the braking point, and then after I went by him, I distanced myself."*

"It is one of the nicest moments of my career, even more satisfying than winning Indianapolis and really gratifying to have so many people pulling for me,"* Andretti said. *"The car remained perfectly balanced throughout the race and the brakes were superb."

"It was not that I was given a break,"* Mario still insists. *"I outbraked him clean. To me, it was just as satisfying a win. Jody tried to say that the only reason I passed him was because the tire was going down; but if that was the case, he would have had a lot of

smoke and a lot of locking up, and there was none of that. So, it was a good, satisfying win."

ANDRETTI WINS L.B. GRAND PRIX

Continued from First Page

third to first in the short dash down Ocean Blvd. before the cars had to make a 90 degree right-hand turn down Linden Ave.

"I don't know if Jody got a hell of a start, or Niki and I fell asleep," said Andretti, "but I thought I had a pretty good 'bits' when I saw Jody go by, already in third gear.

"Then I saw (Carlos) Reutemann coming like a shot and I backed off a bit. It doesn't pay to get your wings clipped right at the start. Reutemann didn't even try to make the turn. He looked like he was flying straight to Acapulco."

Reutemann's suicidal move not only knocked his own Ferrari out of contention but also contributed to the end of world champion James Hunt, Ronnie Peterson in one of the six-wheeled Tyrrells, Hunt's McLaren teammate Jochen Mass and the hapless Vittorio Brambilla. When Reutemann, who started fourth, alongside Scheckter, failed to negotiate the turn his car blocked off Hunt and Peterson and they, too, went down the escape road on Ocean Blvd. Mass and Brambilla, trailing the pack, were caught in the melee and skidded to a stop.

All but Brambilla were able to restart and continue, although Reutemann lasted only eight laps before retiring with suspension problems. It was an embarrassing moment for Brambilla. In the two Grand Prix here he has yet to make it through the first turn of the first lap. Last year he and Reutemann also tangled and Brambilla was left sitting in the middle of the road.

whose promotion has been clouded with unpaid bills from the 1976 race, environmental impact suits that temporarily halted course construction, and, as late as last Thursday, threats of having the race called off because the LBGPA was $150,000 short in purse and transportation money.

"The important thing is we held the race, and it was a thriller," said Pook.

If there is another Long Beach Grand Prix, and that is a moot point, Hunt feels the start should be in a different position.

"It is far too crowded a place to start a race," said the outspoken Englishman. "The organizers should give consideration to starting at a different place, perhaps along Shoreline. You can't turn 24 drivers loose in cars with 500 horsepower in a place with that narrow a turn."

After the first turn the field sorted itself out and it was apparent that Scheckter, Andretti and Lauda were clearly the class, and equally apparent that the Wolf, Lotus and Ferrari were dead even in horsepower and handling.

"When cars are running that equal, it's tough to pass unless the guy ahead of you makes a mistake," said Andretti. "And Jody wasn't making any. I couldn't take any chances trying to do something desperate to pass him, because if it didn't work there was Niki breathing down my neck, just waiting to get past me. It must have been a hell of a race to watch.

"I concentrated so hard I just squeezed everything I could out of the car and it gave me all I needed."

FRUSTRATION FOR HUNT, SCHECKTER

Little Things Got in the Way

BY DWIGHT CHAPIN
Times Staff Writer

James Hunt, who lost his chance to win the Long Beach Grand Prix in a mishap before they got through the first turn Sunday, said it:

"You can live forever on 'ifs.' You can win every race if you do that."

But the words should have come from South African Jody Scheckter, because he did have this race won—for 76 of the 80 laps.

Then a little thing got in the way, a tiny thing that could happen if you were driving a '67 Ford, instead of a space age Formula I machine.

Jody Scheckter hit "a nail or a beer bottle or something" and put a small hole in the center of his right front tire.

ing well when he got to Long Beach and underwent a hospital check. His version is that Pook got the diagnosis—a viral blood infection—and didn't tell him for four days.

"He was gambling with my health," Hunt said, "for the sake of promoting his race. He withheld the information from me. I can't understand why the hospital wouldn't tell me but a third party (Pook). I should have been resting earlier in the week and I wasn't . . . because I didn't know what was wrong with me. Fortunately, I checked again with the hospital a little later and the condition had cleared up."

Pook's version was quite different.

He and a race spokesman—Hank Ives—said Hunt's brother, Peter, had

Post race, Mario Andretti sat in a chair on a platformed rostrum four feet above the ground, surrounded by a sea of newsmen, perhaps as many as 300. He turned in the direction of Team Lotus racing director Colin Chapman and there was a glint in each man's eye. *"Well, I tell you, this is one of the greatest moments of my life, something I will cherish forever,"* Andretti said. *"Colin and I have paid our dues. Now, this is the reward. With Colin alongside me, I feel on top of the world."*

(See the race results on the attached DVD)

Do It in the Streets of Long Beach

DO IT IN THE STREETS — and that's just what they did.

Cars and drivers from around the world raced their ultra light, ultra expensive machines through the streets of Long Beach at speeds up to 190 miles per hour April 3.

Last year's Long Beach Grand Prix did not provide the excitement found in this year's race. Last year, Clay Regazzoni driving a Ferrari led through the race and was never challenged for the position.

This year's race, after a rough start, proved to be exciting to the finish. Mario Andretti and Niki Lauda chased Jody Scheckter around the 13-turn, 2.02 mile course until the 77th lap. There Andretti squeaked around Scheckter closely followed by Lauda. Scheckter finished third. Later it was discovered Scheckter was slowed by a puncture in his right front tire.

World Championship motor racing was started in 1950. The then little known sport had a six race series in Europe. Now, 27 years later, with 18 scheduled races the series is worldwide—including the Long Beach Grand Prix run through the downtown city streets and the United States Grand

Rudolph Caracciola, a German, was the first to take the honor.

By 1950, auto racing popularity had increased and a proper point scoring World Championship was created.

The Formula 1 cars are designer's masterpieces. No space is wasted. The mechanical complexity is awesome. Formula 1 cars are performance demons. They will stop from 60 to 0 miles per hour in less than two seconds (about 80 feet). A passenger car would still be going about 40 miles per hour in 80 feet. The cars will go to 60 and then back to 0 in about four seconds.

Race car wings are simply upside down aircraft wings to produce down rather than lifting force.

With the cars weighing 1265 pounds and engines developing nearly 500 horsepower, it would be like having 400-450 horsepower in a VW Beetle.

The Long Beach Grand Prix course is located in the heart of the downtown section of Long Beach providing a twisting complex racing surface. On Ocean Boulevard, Linden, Shoreline and Pine Avenue, the cars are scrabbling around through the tight corners and straights at close to 80 miles per hour for a timed lap, including 40 miles per hour hairpins and straightaway speeds approaching 190.

The 13-turn facility has undergone some modifications with each race and will be changed yet again with the third race to be held there next year.

Andretti Wins Long Beach Grand Prix

Reutemann Wins Long Beach GP

THIS WEEK

Competition

General

THE UNITED STATES GRAND PRIX WEST

Watch the video on our you tube channel - www.youtube.com/@racinghistoryproject2607 and read the Autoweek and Motor Sport articles on the attached DVD

Losses of over $800,000 had been reported over the past years. Andretti's win turned us *around;* said Jim Michaelian: *"It's no secret that we had suffered some financial difficulty, even though the first two races were successful, but the costs were extraordinary. We were being challenged in terms of getting enough money to continue on. We managed to get the race on, though it took some last minute financial maneuvering to make that happen. But when Mario won and it had that impact of an American driver winning a Formula 1 race in Long Beach, it was a very positive reflection on the city and the event.. It enabled us to go out and get more sponsorships and involvement. That was really the race that turned our financial situation around, and we've maintained that ever since."*

Clay Regazzoni

Niki Lauda

James Hunt

Gunnar Nilsson

Patrick Depailler

Mario Andretti

Emerson Fittipaldi

Jacques Lafitte

Hans Stuck

Vittorio Brambilla

LOEL SCHRADER

L.B. Grand Prix: Sometimes a sticky wicket

Chris Pook has a minute-by-minute schedule for the four days of activity known as the Long Beach Grand Prix.

Everything is supposed to go according to the book.

Naturally, everything doesn't.

But some of the problems Pook has encountered in two races through the streets of downtown Long Beach — a Formula 5000 affair in September, 1975, and the first Formula One race in March, 1976 — exceeded anything he had anticipated.

"Did I tell you about the band?" asks Pook, a transplanted Englishman who is president of the Long Beach Grand Prix Association.

> ***"How in hell does a 100-piece band walk past someone without being noticed?"***

NO, BUT please do.

"Right before the race is to begin, I take a trip around the circuit with Phil Hill (co-race director)," says Pook. "As we were driving up Pine Ave. last March, we heard somebody report on our CB radio that a band was marching down Ocean Blvd.

"Sure enough, as we swung onto Ocean, there was a 100-piece band marching along in formation. I couldn't believe it."

Pook jumped out of the car and yelled to the band director: "Get that band off the track. Some cars are going to be coming through here in a few minutes."

But the band kept right on marching.

"I jumped out of the car and ran up to the director. I said, 'Get the band off the track or

But the bloke kept right on marching along.

"He said, 'I can't stop now. We're due at Ocean and Linden in 10 minutes.'"

BY THIS TIME Pook was coming apart. His schedule had gone to hell but he was determined to stick to it as closely as possible.

"I finally convinced the director he had to get his band off the street. There we were, helping members of the band scramble over the wall and pushing their instruments right after them."

Pook shook his head. "I couldn't believe it."

Later, he checked with a security guard at a gate through which the band must have gained access to the racing circuit.

"I asked him if he'd seen a 100-piece band go through his gate," says Pook. "He looked at me as though I were crazy, so I asked him again.

"Finally, he said to me: 'I thought you were kidding. No, I didn't see a band going through here.'"

So where did the band come from?

"To this day, we don't know how it got onto Ocean Blvd.," says Pook. "How in hell does a 100-piece band walk past someone without being noticed?"

BUT THAT was small stuff compared to the Hermann Goering car incident.

Mercedes-Benz shipped some antique Formula One racing cars from Germany to Long Beach last year for an oldtimers' race involving former racing greats Juan Manual Fangio, Phil Hill, Dan Gurney, Stirling Moss, Maurice Trintignant and others.

"Right after the Mercedes group arrived," says Pook, "they noticed a big van parked

CHRIS POOK...not all beer and skittles

a car that was a replica of one used by Goering when he was one of Adolf Hitler's top military leaders.

"The director of the Mercedes group flew into a rage when he saw the Goering display, and I couldn't blame him. It was an insult to the Germans."

Pook found himself in the middle of an international incident.

"The Mercedes man said they were going to pack up their cars and go home if I didn't get the Goering auto out of there," says Pook. "I didn't know a thing about the Goering display, but I checked and found that our concessionaire had signed a contract that permitted it to be where it was."

cedes people and they agreed to a compromise.

"They would stay if we moved the Goering display to a different area — one that would be out of their sight.

"Then I went to the Goering guy and tried to work out a compromise with him. He was adamant. He had a contract and wouldn't move, and if I did anything about it, he was going to sue.

"We went back and forth until I finally had to make a decision. I told him to get his display out of there or I was going to have it towed away.

"He still refused, so I called a tow truck. When it got there, the bloke caved in. 'All right, I'll move it to a different spot,' he said."

And that is how an interna-

RACING DRIVERS on the Grand Prix circuit lead dangerous lives.

So, understandably, they occasionally go for a romp while away from the track, as demonstrated by Clay Regazzoni of Switzerland and Jean-Pierre Jarier of France a couple of days prior to the 1976 Long Beach Grand Prix.

They had been chauffeured to a party in Beverly Hills with another driver and had consumed a bit of the bubbly.

Emerging from the party, Regazzoni and Jarier found their chauffeur standing by his limousine.

"Let's go," said Regazzoni.

"I can't," said the chauffeur. "I'm still responsible for taking one more of your drivers back to Long Beach."

Regazzoni suggested that the chauffeur go into the home and find the missing passenger.

THE CHAUFFEUR agreed, and left his hat and keys behind on the hood of the car.

Regazzoni and Jarier became impatient when the driver didn't return within a few minutes, so Regazzoni donned the chauffeur's cap and they headed for Long Beach — at a speed considerably above the posted limit.

> ***...and then there was Clay Reggazoni's trophy dash down the freeway, from Beverly Hills to Long Beach, driving a 'borrowed' limousine.***

Upon arriving at their quarters on the Queen Mary, Regazzoni parked the limousine, locked it and, feeling carefree, tossed the keys into the ocean.

The following morning, the limousine owner filed grand theft auto charges against the pair.

But the limo owner was will-

CLAY REGGAZONI
Got tired of waiting

were paid more than $500 for lost time, taxi fare to Long Beach, work by a locksmith and a few other frills, plus more than $400 in tickets to the race.

He was paid off, but costs came off the top of the Grand Prix winner's check.

The winner was Clay Regazzoni.

CHRIS POOK has his minute-by-minute schedule for the 1977 Long Beach Grand Prix, which will be run on Sunday, April 3.

A race of this magnitude, and related events, must go off with precision, particularly since CBS will be televising the main event live to the rest of the nation, with the Southland blacked out.

But Pook isn't going to count on everything going according to the book.

ormula One drivers Mario Andretti, left, and Clay Regazzoni of Switzerland lead the field through the streets of Long Beach, California on April, 3, 1977, during the eginning laps of the Long Beach Grand Prix. (AP Photo)

Flaggers Mary Anne Slick and Pete Watt watch as Hunt goes airborne

1978

The Long Beach Grand Prix West, round four of the Formula One World Championship, was held on April 2, 1978. Long Beach Grand Prix management had discovered how to bring in more spectators; adding additional support races and events and drawing 100,000 fans on race day. This year started off on Thursday with the Prix View Luncheon at the Long Beach Arena, followed by the Concours d'Elegance on Pine Avenue.

Friday included practice, qualifying and a T-Shirt contest at the Edgewater Hyatt House. Saturday brought more practice and qualifying plus a 10,000 meter foot race, a demonstration of the Andretti Mini Grand Prix Cars, the Formula Atlantic Race at 3:30 PM and fireworks at 8:00 PM.

Sunday, raceday, had another Andretti Mini Grand Prix demonstration, the Toyota Celebrity Race at 10:30 AM with the Formula One Race starting at 1:00 PM. The race was broadcast live on CBS but blacked out in Southern California. For the first time, the track announcer was Bruce Flanders.

RACE ROYALTY—The official court for the Long Beach Grand Prix includes Queen Jeannie Sweet, center, and princesses Marsha Holland, left, and Sheri Chiesa. They'll preside over the four days of festivities.

Ain't she Sweet? Lakewood beauty reigns over grand prix

Fireworks show set to music

A fireworks display set to classical an contemporary music, sponsored by th Miller Brewing Co., will kick off activitie at the Long Beach Grand Prix.

The half-hour show, which will begi at 8 p.m. Sunday, will utilize 1,500 shell that will burst in the air in time to th music from "Star Wars," "2001-Spac Odyessy" and the classical works "Th Planets" by Holst, "The William Te Overture" by Rossini and the "181 Overture by Tchaikovsky.

Gordon Johncock won the Toyota Celebrity Race, followed by Dan Gurney and Don Prudhomme. Howdy Holmes won the Formula Atlantic Race followed by Tom Gloy and Kevin Cogan. As for the Formula One race, only 22 cars would make the grid but there were 30 entries, requiring a one hour pre qualifying session on Friday morning for the eight non FICA members. From these eight, the fastest four would join the rest of the field

in qualifying for one of the 22 starting spots. The start had been moved from in front of the pits on Ocean Boulevard to the curving straight on Shoreline Drive in order to avoid another first corner tangle. The strategy seemed to work, as everyone got through cleanly, though John Watson's late braking maneuver down the inside caused him to exit wide from the hairpin and force polesitter Reutemann wide with him, so Villeneuve tucked into the lead as they exited. Reutemann took the lead on lap 39, just before the halfway point, Villeneuve came up to lap Clay Regazzoni, who was in a battle with Jean Pierre Jabouille's Renault. Rather than wait until the straight, the Canadian tried to get by in the twisty section leading up to Ocean Boulevard. Regazzoni braked earlier than Villeneuve expected, the Ferrari's right front wheel hit the left rear of the Shadow and was launched over the white car into the wall.

This left Reutemann in the lead, ahead of Jones, then a long gap back to Andretti. Lap after lap, Jones hounded Reutemann, but the Ferrari was too fast down the straight for him to get by. On about lap 47, with Jones still on the Ferrari's tail, the front wings of the Williams strangely flopped down, the result of a fabrication failure. Jones continued to battle, but began losing around a second a lap. Then his fuel pressure began fluctuating, the engine sputtered and, sometimes cut out entirely. As one car after another passed, the struggling Williams fell all the way to eighth place. Recording the fastest lap of the race on lap 27 and repassing Emerson Fittipaldi on the last lap for seventh place were his only consolations after a spectacular drive. Reutemann went on to score the victory, collecting $52.357 for the win. 11 seconds back was Mario Andretti with Patrick Depailler in third.

Carlos Reutemann

(See the race results on the attached DVD)

Reutemann rules Long Beach

By The Associated Press

For Carlos Reutemann, the handsome and moody Argentine driver, it seems it's all or nothing at all.

The 34-year-old Formula 1 veteran and his red Ferrari either leave the opposition thrashing helplessly in their wake, or both wind up on the sidelines long before the race is over.

Reutemann overwhelmed everyone at the Brazilian Grand Prix earlier this season, and took one of the easiest victories of his career. But at Argentina and South Africa, he got nothing.

It was feast time again Sunday, though, as Reutemann started on the pole position, led the last half of the race and took an easy victory over Mario Andretti in the U.S. Grand Prix West.

"To be consistent is what we need most," said Reutemann, recalling that he got off to a similar start last season only to drop completely from contention in the second half of the season because of spotty results.

But in 1977, Long Beach was the beginning of the end for Reutemann, who dropped out of the race very early.

Ferrari has won the pole position in each of the three years of the Long Beach race, and won the race two out of three.

Last year, Reutemann came to Long Beach leading the World Championship standings, this year he leaves with a share of the lead.

Reutemann and Andretti are tied with 18 points apiece after the fourth of 16 races this season. Defending champion Niki Lauda of Austria slipped to fifth behind Patrick Depailler of France and Andretti's teammate Ronnie Peterson of Sweden.

Reutemann had a comfortable 11-second margin at the end of the 80 laps, averaging a record 87.069 miles per hour. Only half of the starting field of 22 finished; the 2.02-mile course through the downtown streets of Long Beach is reputedly very tough on axles, transmissions and suspensions.

Behind Andretti, there were only three other cars on the same lap, and even those were hopelessly far behind.

A record crowd of more than 100,000 watched, some 75,000 paid, according to organizers.

Reutemann's share of the $400,000 purse, behind only the Indianapolis 500 and Daytona 500 in this country for total dollars, was not announced. All Formula 1 prize money is distributed after the season by the Formula 1 Constructors Association. The payout is never announced to the public.

U.S. GRAND PRIX WEST (80 laps) — 1. Carlos Reutemann (Argentina), Ferrari, 80 laps; 2. Mario Andretti (Nazareth, Pa.), Lotus, 80; 3. Patrick Depailler (France), Tyrrell, 80; 4. Ronnie Peterson (Sweden), Lotus, 80; 5. Jacques Laffite (France), Matra, 80; 6. Riccardo Patrese (Italy), Arrows, 79; 7. Alan Jones (Australia), Williams, 79; 8. Emerson Fittipaldi (Brazil), Copersucar, 79; 9. Rolf Stommelen (Germany), Arrows, 79; 10. Clay Regazzoni (Switzerland), Shadow, 79; 11. Jean Pierre Jarier (France), ATS, 75; 12. Patrick Tambay (France), McLaren, 74.

DID NOT FINISH — 13. Jody Scheckter (South Africa), Wolf, 59, front wing damage; 14. Vittorio Brambilla (Italy), Surtees, 50, engine failure; 15. Jean Pierre Jabouille (France), Renault, 43, rear end tire; 16. Gilles Villeneuve (Canada), Ferrari, 38, accident; 17. Niki Lauda (Austria), Brabham, 27, electrical failure; 18. Didier Pironi (France), Tyrell, 25, gear box; 19. Arturo Merzario (Italy), Merzario, 17, left front suspension; 20. Jochen Mass (Germany), ATS, 11, oil line leak; 21. John Watson (North Ireland), Brabham, 9, electrical; 22. James Hunt (England), McLaren, 5, broken front wheel.

TOYOTA CELEBRITY RACE (10 laps) — 1. Gordon Johncock, 58.529 m.p.h.; 2. Dan Gurney, 58.510; 3. Don Prudhomme, 58.060; 4. James Brolin, 57.744; 5. Kent McCord, 56.881; 6. Bill Simpson, 57.798; 7. Paul Williams, 55.161; 8. Kitty O'Neil, 54.736; 9. William Shatner.

Street fighters

The world's top Formula 1 race car drivers hit the streets of Long Beach Sunday for the annual running of the United States Grand Prix West. Top, Mario Andretti, who finished second in the race, gets the thumbs-up sign from a pit crew member. Two other pit crew workers, above, work on their machine while, below, Jody Scheckter gives his car fuel throttle down a straightaway. (News Chronicle photos by Steve Kilgore)

It's a Big Day for Reutemann and Long Beach

Los Angeles Times
Sports
BUSINESS & FINANCE
CC PART III †
MONDAY, APRIL 3, 1978

More Than 100,000 See Argentine Score Easy Win in Ferrari

BY SHAV GLICK
Times Staff Writer

Carlos Reutemann, Ferrari and Michelin won the third U.S. Grand Prix West through the streets of downtown Long Beach Sunday to bring the world Formula One season into a deadlock after four of the 16 races.

It was Reutemann's second win and gave him 18 points and a tie with Mario Andretti, second place finisher Sunday, for the world driving championship. It was the second win for Ferrari against two for Lotus and the second win for Michelin tires against two for Goodyear.

But the biggest winners were promoter Chris Pook, the Long Beach Grand Prix Assn. and the city of Long Beach. Pook announced that 75,000 tickets were sold—more than the total viewers at the previous races—which meant more than 100,000 watched the race that was designed to confirm Long Beach's claim to be the International City.

Reutemann was an easy winner in the 80½-lap race around the 2.02-mile circuit, taking the checkered flag in his red-and-white Ferrari more than 11 seconds ahead of Andretti. Although the Argentine was not challenged through the final 25 laps, the earlier stages of the race were more exciting than any seen before in Long Beach.

Gilles Villeneuve, Reutemann's 26-year-old teammate from Canada only one year out of Formula Atlantic racing, jumped his more illustrious competitors on the first turn, the Queen's Hairpin, and drove brilliantly in front for 38 laps before succumbing to youthful enthusiasm.

Coming up on 1976 Long Beach winner Clay Regazzoni in a chicane through the parking lots on the approach to the Pine St. hill, Villeneuve picked a tight place to make a pass. As he swept past Regazzoni, the veteran Swiss driver shut the door on him and his Shadow caught a wheel on the Ferrari. The sudden impact swung Villeneuve around and sent him crashing backwards into a barrier of tires—and out of the race.

"It is not the type of place where Clay would expect someone to overtake him," said Reutemann of his teammate's inexperienced move.

Regazzoni, however, found himself involved in several other incidents during the long and difficult race on a magnificent spring Sunday.

Alan Jones, the chubby Australian driving a Williams sponsored by Saudi Arabian money, was clipping off the day's fastest laps while moving up from eighth to second in 40 laps. His trip around the course on Lap 27 at 88.65 m.p.h. was the quickest of the day by anyone.

Jones appeared capable of catching Reutemann, who had inherited the lead from Villeneuve when the Canadian crashed. However, in an attempt to get past Regazzoni he drove into the rear end of Regga's car and bent his front spoilers (wings). This, coupled with gear box problems that developed later in the race caused Jones to drop back and eventually

Please Turn to Page 7, Col. 1

Mario Andretti

James Hunt

Patrick Depailler

John Watson

Patrick Tambay

Didier Pironi

Ronnie Petersen

Niki Lauda

Gilles Villeneuve

Emerson Fittipaldi

Autoweek

Reutemann Wins Long Beach GP

THIS WEEK

Competition

General

The United States Grand Prix West

Watch the video on our you tube channel - www.youtube.com/@racinghistoryproject2607 and read the Autoweek and Motor Sport magazine article on the attached DVD

1979

The 1979 United States Grand Prix West, held on April 8, 1979, now had a sponsor, allegedly for $300,000 and was formally titled the Lubri Lon Long Beach Grand Prix. The crowd was estimated at 81,000.

The week got busy. On Tuesday, there was a benefit art show featuring grand prix scenes at Harbor Bank. Thursday was the usual Prix View Luncheon and Pine Avenue Concours. Friday was practice, a new Grand Prix Motorsports Expo in the arena and a celebrity disco kickoff party at the Elks Club; admission was $15.

Saturday was more practice, qualifying, a 10,000 meter foot race sponsored by Nike, the Toyota Celebrity Race, televised live on KNXT-TV and the Formula Atlantic Race with fireworks at 8:00 PM. Sunday, the AFX Aurora Pepsi Slot Car Challenge ran in the garage and the Formula One Race started at 1:00 PM. CBS televised one hour of qualifying on Saturday and over two hours on Sunday, blacked out in Southern California.

Saturday's Toyota Celebrity Race was won by Bruce Jenner who was also the celebrity class winner with Al Unser, the professional class winner, in second and Don Prudhomme in third. Tom Gloy won the Formula Atlantic Race followed by Kevin Cogan and Bill Brack.

In Formula One, Canadian Gilles Villeneuve captured the pole, fastest lap and the win for Scuderia Ferrari, followed by his teammate Jody Scheckter, as Ferrari took a big step toward reclaiming the Constructors' and Drivers' Championships from Team Lotus.

Villeneuve's win came by almost half a minute over Scheckter, and Alan Jones joined them on the podium for Williams. It was the third win of Villeneuve's career, his second in succession, and the third United States Grand Prix win in a row for Ferrari.

Staff photo by Bill Camerc

Gilles Villeneuve enroute to winning pole position for today's Long Beach Grand Prix.

Qualifying was a battle between Ferrari, Lotus and Ligier and the circuit was littered with broken cars by the end of each session. Carlos Reutemann, in the second Lotus, held the pole until the very end of the final session, when Villeneuve bumped him.

As with the previous year, the start would be on the Shoreline Drive straight, rather than in front of the pits, giving the drivers more of a run down to the first corner. On the warmup lap preceding the drive around to the grid, Reutemann's engine cut out, and he had to be helped back to the pits. The problem was quickly solved, and, while the organizers ruled that he must start from the back of the grid, Reutemann ignored the officials and went back out.

The cars made their way around for the start on Shoreline Drive, and, as they approached the grid markings, pole sitter Villeneuve drove right past his starting position. This confused the entire field, who were taking their cue from Villeneuve. Some actually thought the race had started, and everyone ended up back at the pits. On the way to the grid, Laffite's Ligier had suddenly slid across the track when the back end seized. He was allowed to switch to his spare car and, when they finally got under way, he started the race from the pits with Reutemann. Laffite retired with overheated brakes, and Reutemann broke a driveshaft so two of the top five qualifiers were out early, and the door was wide open for the Ferraris.

Villeneuve got to the first corner ahead of Scheckter and quickly began to draw away. Scheckter lost a position to Depailler on the first half lap, but began pushing hard to take back second place. When Depailler missed a gear, Scheckter nearly hit him in the back, and Jarier slipped his Tyrrell past both of them. On the next lap, Scheckter also got around Depailler, whose troubles with fourth gear would last the entire race.

Villeneuve continued to run untroubled in the lead, expanding his margin whenever he chose and setting the race's fastest lap before half distance. Villeneuve made $77,127 for the win.

(All the race results are on the attached DVD)

Watch the video on our you tube channel - www.youtube.com/@racinghistoryproject2607 and read the Motor Sport magazine article on the attached DVD

Villanueve wins L.B. Grand Prix, loses $6,000

By Mike Morrow
Staff writer

NEWS-PILOT
Sports
Mon., April 9, 1979 B2

Salud, El Commandante.

Glasses of champagne were lifted Sunday night in tribute to 81-year-old Enzo Ferrari, auto racing's supreme commander.

For the third time in four tries, a fire engine red Ferrari was in the winner's circle at the Long Beach Grand Prix. Young French Canadian Gilles Villeneuve held aloft the silver cup symbolic of victory.

Starting on the inside pole position, the 5-foot-7, 140-pound Villeneuve broke ahead and held onto the lead for the entire 80-lap race, beating teammate Jody Scheckter by a commanding 29.38 seconds.

It was the second straight one-two finish

process.

At the back of the field, Reutemann got his Lotus going, but it didn't go very long, coming out of the race after 18 laps.

The Ferraris figured to be pretty close to the lead anyway, with just about everyone giving them the advantage following their impressive win at Kyalami, South Africa, last March 3.

Rene Arnoux

Patrick Tambay

Hans Stuck

Jody Scheckter

Lubri
Beach
Marlboro

LONG
GRAND

Carlos Reutemann

Didier Pironi

Nelson Piquet

Jean Pierre Jabouille

Niki Lauda

Mario Andretti

Grand Prix Here to Stay, Finally

BY TOM HAMILTON
Times Staff Writer

ANAHEIM — Chris Pook, the president of the Long Beach Grand Prix Assn., can remember when he had problems getting Long Beach city officials to agree on anything.

Pook had a brainchild in 1975 — racing through the streets of Long Beach, but faced more obstacles trying to get the race off the ground than the drivers eventually did manuevering through the 2.02-mile course.

"It's funny how many city officials are now saying, 'This sure was a great idea I had,' " said Pook, as he prepares for the fifth race that will run through the downtown streets of the city that was once known as "Omaha, Neb. West."

In 1975, Long Beach was a city looking for publicity. The sleepy Pacific Coast city with the sixth-largest population in the state had an identity crisis. It got lost somewhere between Los Angeles and Disneyland.

Enter Pook and his race idea. Not exactly fitting with the Queen Mary, the Pike and the senior citizens of Long Beach. But an idea to get international coverage in newspapers across the nation as well as the world.

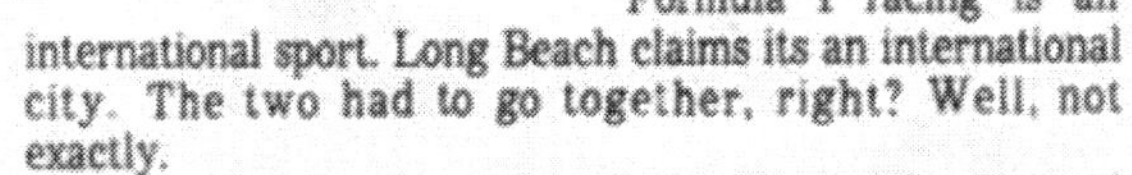

Chris Pook

Formula I racing is an international sport. Long Beach claims its an international city. The two had to go together, right? Well, not exactly.

Pook confronted problems ranging from the Coastal Commission to the high-price tag from the Formula I Constructors Assn. just to get the drivers and cars to Long Beach. The initial Formula I cost $530,000 to stage before the show even started.

Pook is now in the final stages of preparation for the fourth running of the Long Beach Grand Prix Formula I race on April 6-8. And he says the city and the public have now accepted the race after a long struggle.

"It's a sophisticated market in Southern California," said Pook.

"It took three or four years to get accepted," he added. "We've managed to hang on by our fingernails a couple of years and now we are established."

Part of being established is evident by obtaining a major sponsor for the race for the first time — Lubri Lon, a metal treatment that coats the internal lubricated surfaces of an engine. Lubri Lon paid over $300,000 for the sponsorship, according to Pook.

Then there's television coverage. CBS will air one-hour of qualifying on Saturday and two hours and 20 minutes on Sunday (delayed until later Sunday evening in Southern California).

But the real leverage of any sporting event is ticket sales. Pook says ticket sales have doubled last year's totals with three weeks remaining before the race.

"Last year, we had 77,500 fans on race day and a total of

the South African Grand Prix this year, the course record was smashed by eight seconds.

"Drivers were entering the end of the straightaway there at 180 mph and braking 200 meters for the turn last year. This year, drivers were still going 180 mph but braking only 100 meters before the turn."

Pook initially signed a 10-year contract with the city of Long Beach in 1975 to stage the event. He says the Long Beach Grand Prix Assn. will probably negotiate a new, 10-year pact after next month's race.

"We've had a lot of different cities approach us about the race," Pook said. "Las Vegas, San Francisco, San Diego and Miami have all expressed interest in staging the race."

Formula I racing in the U.S. is indeed lucrative. Sixteen races are held throughout the world and two of them have been granted to the U.S. — one at Long Beach and the other at Watkins Glen in New York.

"Street courses make good spectator sport — you rarely have a chap that runs away with a race and 65% of the cars finish," said Pook. "Long Beach is a slow, safe course with an average speed on only 84 mph so the fans can actually see what's going on in a drivers' cockpit."

Spectators aren't the only ones interested in the Long Beach Grand Prix. Race officials receive thousands of media crediental requests from all over the world.

"People love to come to Long Beach," said Pook. "It's become known along the circuit as a very hospitable race. We could probably fill a DC-10 with the media who travel to the race from across the world."

This year's race didn't come off all that smoothly. Pook was fined $20,000 by the Formula I Constructors Assn. for moving the race back a week because it conflicted with another race in Europe.

Does he forsee other problems before April 8?

"I suppose the only problem that still confronts us is the politics involved in staging a major race through the streets of a major city," he said. "Sometimes I have to jump on people's desks just to get their attention.

"But remember, the biggest automobile market in the world is right here in Southern California. The auto industry wants this race established in Long Beach."

ND
FREEWAY

Haan,Inc.
GEORGE FOLLMER
4
Tyrrell
6
parmalat

1980

The 1989 United States Grand Prix West, held on March 30, 1980, was now the Toyota Long Beach Grand Prix, round four of the Formula One season. Toyota signed a three year, contract, reputed to be $850,000,, to sponsor the event for the next three years. 182,500 people attended over three days.

The schedule had expanded to start the previous Saturday with the Charity Concours d'Elegance and Charity Chili Cook Off in the Queen Mary parking lot. Tuesday was the charity art show at Harbor Bank, Wednesday, the Prix View Luncheon at the Convention Center and Thursday, the Pine Avenue Concours and Race Car Show.

Racing got underway on Friday with practice for Formula One, historic Formula One, the Toyota Celebrity Race and Formula Atlantic. The Motorsports Expo in the arena also opened for the weekend. More practice and qualifying took place on Saturday along with the Bell Helmet's sponsored bicycle race at 2:15 PM, the historic Formula One Race for the amateur drivers at 3:10 PM and the Formula Atlantic Race at 3:45 PM.

Sunday, race day, featured a Toyota owners drive around the track, the historic Formula One Race, this with professional drivers, at 11:55 AM, the Nike 10,000 meter foot race at 12:35 PM, the Formula One drivers parade at 1:10PM with the big race starting at 2:00 PM.

For the benefit of the TV cameras, the porno theaters at start / finish were covered with banners that read "Join Gunnar's Fight Against Cancer".

TOYOTA GRAND PRIX OF LONG BEACH

WORLD CHAMP JODY SCHECKTER AND FRIENDS THEY'RE ALL RACING THRU THE STREETS OF LONG BEACH MARCH 30, 1980 in the TOYOTA GRAND PRIX of LONG BEACH

The World of Racing's Greatest Value!

March 28-29 action includes Grand Prix Qualifying, TOYOTA Pro/Celebrity Races, Historic Grand Prix for Vintage Cars, Formula Atlantic Championship, Bell Helmets' Bicycle Classic and much More. March 30 more Historic Cars, Blue Angels, Nike Footrace and the 5th Grand Prix through the streets of Long Beach.

For Ticket Brochure
Clip and Mail Today

To: Toyota Grand Prix of Long Beach
100 East Ocean Blvd., Suite 908-T2
Long Beach, CA 90802

Name

Address

City State Zip

Phone Ticket Hotline
(213) 436-9953
For Info or Ticket Purchase
Tickets also available through TICKETRON and BASS Ticket Outlets New LBGP Box Office: 202 East Ocean, Long Beach (front of Breakers Hotel)

SEE THE GRAND PRIX ON DISPLAY DEL AMO SHOPPING CTR. FEB. 22-23-24

BE FIRST ON THE COURSE

SUNDAY MORNING, MARCH 30, TAKE A PARADE LAP AROUND THE GRAND PRIX COURSE IN YOUR TOYOTA.

Here's a unique chance for Toyota drivers to get the feel of the Grand Prix Race course, hear the crowds and sense the excitement of Grand Prix Racing through the streets of Long Beach.

The parade lap is just part of a special Toyota Rally Package. The package includes the parade lap, close-in parking for the race, ear plugs, T-shirt, Grand Prix Button, stroker cap and informative pre-race talks by a formula one driver and a Toyota Celebrity Driver. All licensed drivers are welcome—but only the first 250 can be accommodated.

See your Toyota dealer for details on the Rally and for a close look at the official pace car of the Toyota Grand Prix of Long Beach, the remarkable Celica Supra. Then Sunday morning, the day of the race, weather permitting, be first on the course at the Toyota Grand Prix of Long Beach.

TOYOTA OH WHAT A FEELING

The professional historic Formula One Race was won by Carroll Shelby, with Dan Gurney second and Jack Brabham third. The Toyota Celebrity Race was won by Parnelli Jones, also winning the professional class. Second and winning the amateur class was Gene Hackman with Bruce Jenner in third. The Formula Atlantic Race was won by Tom Gloy with Kevin Cogan second and Bob Earl third.

Long Beach Grand Prix finds sponsor

By JIM SCHULTE
Sun Sports Writer

LONG BEACH — The Long Beach Grand Prix was assured three more years of operation Thursday when it was announced that Toyota Motor Sales (USA) would be the new major sponsor of the Formula One race, beginning with the 1980 event and continuing through 1982.

The Japanese auto maker is already sponsoring the U.S. Grand Prix at Watkins Glen, N.Y., and has been

A race day crowd of 85,000 saw Nelson Piquet qualify on the pole , lead every lap and win by 49 seconds, collecting $86,292. Riccardo Patrese was second driving an Arrows A3. And third was Emerson Fittipaldi driving a Fittipaldi F7.

The race would be the last in Formula One for Swiss driver Clay Regazzoni, who suffered spinal damage when he lost the brakes in his Ensign and struck a parked car followed by a concrete barrier at 150 mph.

James Hunt nearly made a comeback with McLaren, asking for $1 million for the race. This opportunity came about when French rookie driver Alain Prost broke his wrist during practice for the South African Grand Prix, and was not fully fit to drive at Long Beach. The team's main sponsor, Marlboro, offered half the figure but negotiations ended after Hunt broke his leg while skiing. Stephen South substituted but South failed to qualify.

A sidebar to this; Rick Mears was denied a race entry as he had not entered three months in advance as required by FIA.

The only misstep of Piquet's weekend came in the Sunday morning warmup when he and Derek Daly collided entering the corner after the pits. The Brabham was vaulted into the air and landed heavily on all four tires, while Daly's Tyrrell went down the escape road. After careful examination, the Brabham crew decided that the car was fine and ready to race.

Piquet wins pole for U.S. Grand Prix-West

LONG BEACH, Cal. (AP) — Nelson Piquet blew away the year-old qualifying record for the U.S.-West Grand Prix yesterday, winning the pole for today's race with an average 93.598 m.p.h. in his Brabham.

The 27-year-old Brazilian toured the 2.02-mile course through the streets of Long Beach in one minute, 17.694 seconds during the second and final hour of qualifying.

It was the third time in the session Piquet eclipsed the record 92.255 m.p.h. set in a Ferrari last year by defending race champion Gilles Villeneuve of Canada.

The top four qualifiers broke the old mark. Rene Arnoux of France put his turbocharged Renault on the front row with a lap of 92.414. Patrick Depailler of France in an Alfa Romeo and surprising Jan Lammers of Holland in an ATS grabbed second-row spots at 92.379 and 92.304, respectively.

Acknowledging that cars up front have a definite advantage on the tight Long Beach circuit, Piquet said, "I have a good chance, but anyone can win the race. The thing is, the car ran very well with a full tank."

For the first time in the five years the Long Beach race has been run, there was no Ferrari on the front row. The Italian team won three of the first four races. Villeneuve qualified 10th, defending world champ Jody Scheckter of South Africa 16th.

American and 1978 world champion Mario Andretti qualified his Lotus 15th at 91.170.

(See the complete race results on the attached DVD)

UPI

Long Beach Grand Prix

Piquet easy winner

LONG BEACH, Calif. (UPI) — If Nelson Piquet of Brazil had known it was going to be so easy, he might have brought the wife and kids along for the ride.

The only trouble the 27-year-old Piquet had Sunday was during the pre - race warmups, when he spun his car once and then had a minor collision with another car as he raced to the wire - to - wire victory in the Long Beach Grand Prix. He finished nearly 50 seconds ahead of Riccardo Patrese of Italy.

The race was marred by a crash in which Clay Regazzoni of Switzerland sustained a spinal in- Ensign-4 crashed into a concrete retaining wall along city streets at 150 mph. The driver was trapped inside his car for 20 minutes and then rushed to nearby St. Mary's Hospital where he was reported in stable condition.

Piquet, winless in 23 Grand Prix races prior to Sunday's triumph, started from the pole position and was never headed, as he pulled away in his Brabham-Ford and 14 of the 24 drivers were knocked out of the race. It was the first win for a Brabham since 1978 and the first time the car had finished a Long Beach race in four years.

80 laps over the 2.02 - mile, 12 - turn Long Beach course in 1 hour, 50 minutes and 18 seconds for an average speed of 88.476 mph. "Then, after (Alan) Jones stopped, I relaxed even more."

Piquet, who finished second in the Argentina Grand Prix earlier this year, opened up a lead of 53.25 seconds over Patrese by the 55th lap and then eased to the checkered flag.

The win moved him into a tie with Rene Arnoux of France in the world championship point totals with 18 each. Jones, who was running se-

THE UNITED STATES GP (WEST)

Watch the video on our you tube channel - www.youtube.com/@racinghistoryproject2607
and read theGrand Prix International and Motor Sport articles on the attached DVD

RECORD GRAND PRIX TURNOUT

182,500 Race Fans Invade Privacy Of Harry the Tramp in Long Beach

By **ROBERT J. GORE**
Times Staff Writer

LONG BEACH — "Varoom, varoom, vaaroooom . . . bah!" shouted Harry the tramp as he sat on a blue plastic milk crate, peering through a crack in the fence at the Grand Prix cars howling past.

Harry, in spite of his professed hostility toward the fifth annual race at Monte Carlo West, was watching intently; somehow managing to keep track of the leaders on a tattered newspaper listing.

A record crowd joined Harry, jamming around the serpentine course to enjoy one of two Formula One street contests in the world. (The other is Monte Carlo.)

About 110,000 fans came to Long Beach — home of the Grand Prix, $11 T-shirt, $10 parking space and $1 ice cream sandwich.

There were 182,500 paid admissions for the three days of racing, said Chris Pook, president of the Long Beach Grand Prix Assn. This is 22,000 more than last year.

Pook said the preliminary financial report would be available in three weeks. In its first year, the race lost $1.2 million, but this was turned around into a profit of $307,065 in 1979.

"We had our act together a little better this year," Pook said. "We're very happy with the turn out and the race went far better than in the past."

Tinted varying degrees of pink by a summer sun, the spectators could see the drivers fighting their steering wheels, gear shifts and gravity in an intense effort to maintain control of the hurtling $150,000 cars.

The race was marred by a serious accident. Swiss driver Clay Regazzoni apparently lost his brakes at the end of the longest straight stretch on Shoreline Drive.

Regazzoni's car, entering the tightest hairpin turn, left the course, taking an emergency exit road. Traveling about 120 miles per hour, his car clipped the rear of a parked racer.

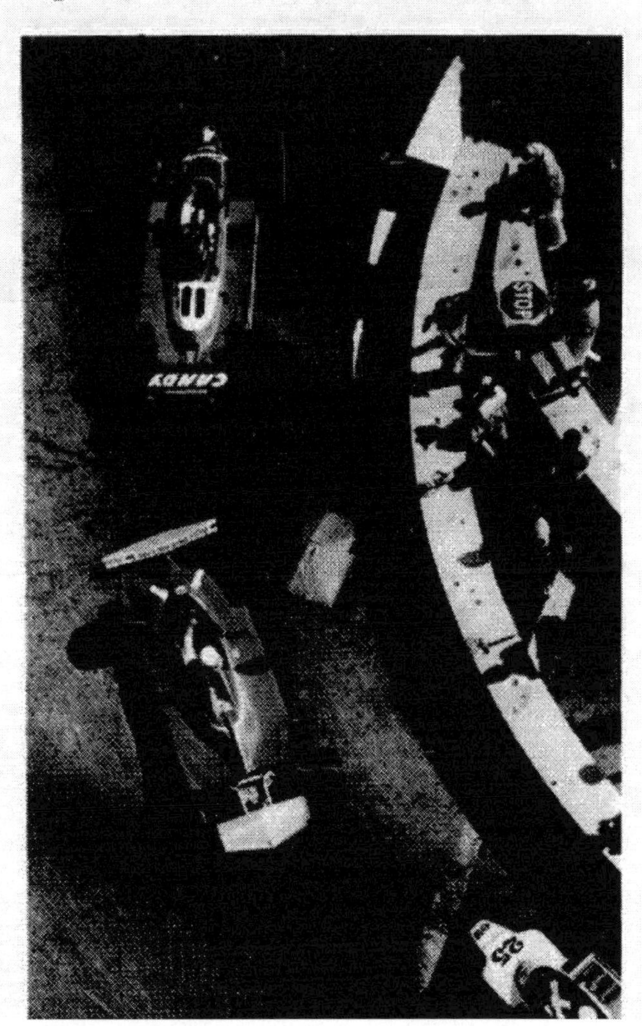

DUEL IN THE MUD—Known to their fans as Pretentious Pamela, waving to crowd, and Double Delicious Dede, below, mud wrestlers perform on Queen Mary in Grand Prix festivities Sunday.

Times photos by Thomas Kelsey

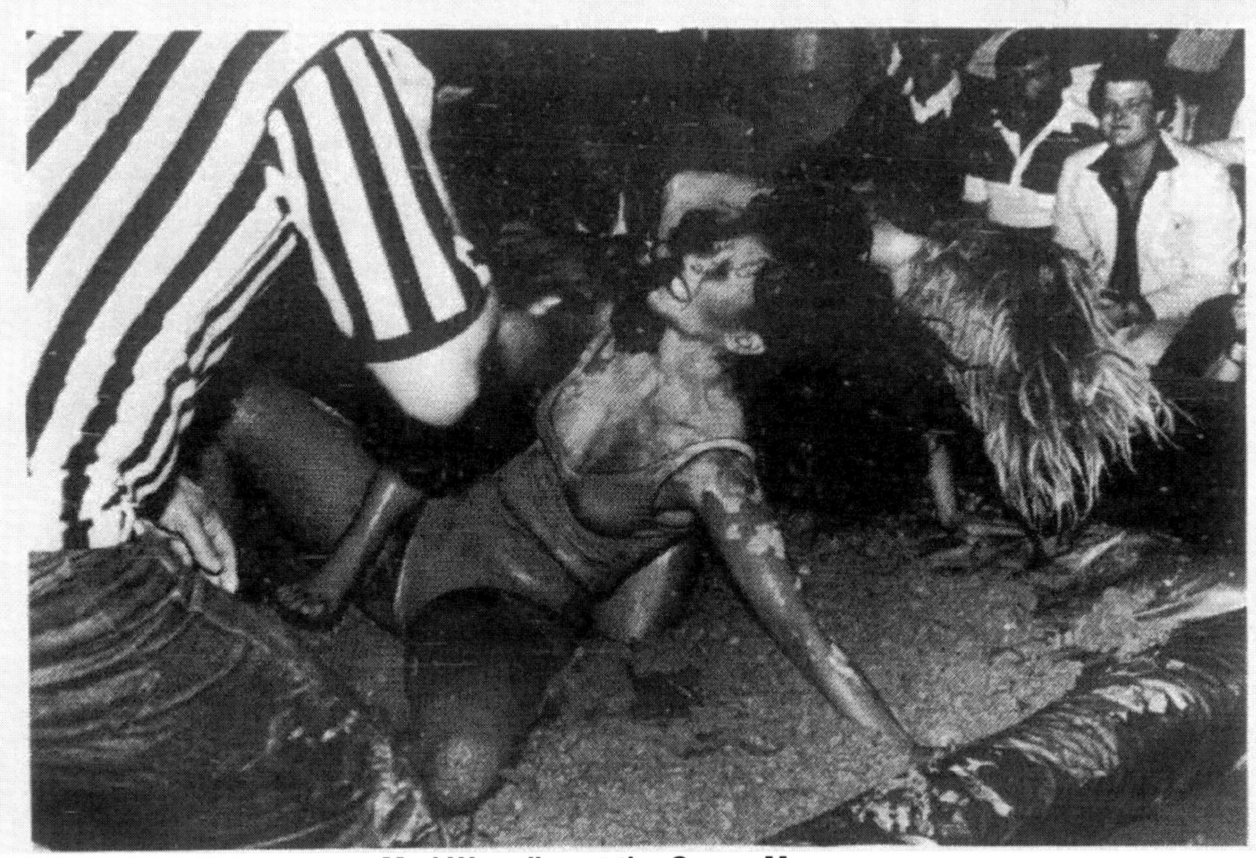

Mud Wrestling at the Queen Mary

Nelson Piquet

Geoff Lees

Mario Andretti

Elio de Angelis

Emerson Fittipaldi

Ricardo Patrese

Jan Lammers

Clay Regazzoni

Brazilian breezes in crash-marred Grand Prix

NEWS-PILOT

Sports

Mon., March 31, 1980 A10

Former stars rap new cars

Not everyone shared the excitement of Sunday's Long Beach Grand Prix.

"The cars today look a little like slot cars," said Carroll Shelby, the one-time driving champion who turned auto racing into a multi-million dollar business.

"Without being negative, it is hard for me to say much about today's racing," said Dan Gurney.

"My thoughts are jaded, I like the good old days," said Phil Hill.

Shelby, Gurney and Sir Jack Brabham finished 1-2-3 Sunday in a five-lap, historic car race, covering 10.1 miles in a time of 9 minutes, 45.7 seconds, with the winner averaging 62.083 miles per hour.

It wasn't this race that brought the 100,000 fans to Long Beach, though.

"We are slower, but we have a lot more fun," said Shelby.

"I don't think the drivers of today are having fun, it has become too serious, too expensive," said Brabham.

"I would like to see them return to smaller engines, narrow tires, see more slipping and sliding," said Gurney.

Seven drivers competed in the event, named after the late Peter DePaolo, the first man to ever win the Indianapolis 500 at an average speed of 100 miles per hour.

Shelby, driving an Aston-Martin, took the lead at the start, with Gurney and Brabham running second and third.

The field included two Aston-Martins, a Maserati 300S, a Cooper, Ferrari 250TR, Talbot-Lago and Honda.

"I would like to see them slow the cars down, make it

Staff photo Michael Justice

Nelson Piquet (5), above, closes in on Didier Pironi (25) early in Sunday's running of the Long Beach Grand Prix. Piquet passed Pironi on the next turn and went on to win the race. Below, a member of the fire rescue team runs toward the wreck of the car driven by Clay Regazzoni, who was seriously injured when his car disintegrated on impact and threw him into the rubber tire barrier at right.

Piquet's first win easy; one driver injured

By Mike Morrow
Staff writer

Nelson Piquet was at a loss for words when a microphone was stuck in his face as he took the winner's stand Sunday afternoon at the Long Beach Grand Prix.

"I have never done this before," he said.

But Piquet, a non-winner in two years of championship Formula One race car driving, turned in an easy victory, taking advantage of an absence of any challenger for more than half the 81 laps.

Alan Jones, his closest challenger and a Palos Verdes Peninsula resident, was forced off the course in the 47th lap, giving the leader a 40-second lead over Riccardo Patrese.

From that point, Piquet said later, "it was only a matter of keeping concentration."

"I feel great. I have no words to explain how I feel," said Piquet, a 27-year-old Brazilian who gained the pole position Saturday and moved right in front at the outset, never looking back.

At the back, though, was the most spectacular event of the race, a crash involving veteran driver Clay Regazzoni.

Regazzoni apparently lost braking compound as he was going into the Queen's Hairpin turn.

Unable to brake at 160-170 miles per hour, he had to steer his car off the course, crashing it into the retaining area down an escape road.

He hit the fence in the air, bouncing off and landing hard, destroying his Ensign MN 180.

Safety crew personnel were unable to

There's More to Long Beach Than a Grand Prix

By SHAV GLICK

MOTOR RACING

[illegible]

Rick Mears Denied Entry in Long Beach Grand Prix

From a Times Staff Writer

LONG BEACH—Indianapolis 500 champion Rick Mears was denied permission Wednesday to race in Sunday's Toyota Grand Prix of Long Beach by the sanctioning Federation International Sport Automobile.

Yves Leon, secretary of FISA, wired Chris Pook, president of the Long Beach Grand Prix Assn., that Mears' on-

Actor Gene Hackman and former Indy winner Parnelli Jones celebrate their victories in 1980's Toyota/Pro Celebrity Race. The Grand Prix queen and E.A. Hagan of Toyota Motor sales look on.

1981

The 1981 United States Grand Prix West, held on March 15, 1981, was now officially the Toyota Grand Prix of Long Beach. This was the first race of the 1981 Formula One season. KLAC radio had coverage of the race week on the "Pit Row Show" Monday through Friday and complete coverage of the race on Sunday. CBS, with 16 cameras this year and sportscasters Ken Squier, David Hobbs, Brock Yates and Barrie Gill, covered the race live but it was delayed in Southern California until 11:45 PM. Thursday was the Prix View Luncheon and Pine Avenue Concours. The Grand Prix of Long Beach Association took on some new ventures, some good, some not so good, in managing the Las Vegas Formula One Race and CART at the Meadowlands and Dallas and IMSA at Del Mar..

Friday was practice and qualifying plus a Grand Prix Kickoff Party at 7:00 PM at the Queen Mary. Saturday was more practice and qualifying with the Toyota Celebrity Race at 11:45 AM, the Bell Helmets Bicycle Race at 2:10 PM, the Formula Atlantic Race at 4:15 PM and another Grand Prix Kickoff Party at 7:00 PM at the Queen Mary.

Sunday had a Toyota Owner's Rally and track drive at 9:45 AM, then a 10K run sponsored by Nike at 10:15 AM. New on the schedule this year was a ten lap Pro Kart Challenge featuring high horsepower go karts at 11:50 AM on Sunday, a ten lap motorcycle sidecar race at 12:20 PM followed by the Formula One Race.

Larry Coleman won the sidecar race while Geoff Brabham won the Formula Atlantic race followed by Jacques Villeneuve and Danny Sullivan. The Toyota Celebrity Race was won by Robert Hays, who also won the amateur class followed by professionals Elio de Angelis and Dan Gurney

82,000 paid attendees plus an estimated 25,000 watching from roof tops and balconies saw defending World Champion Alan Jones finished nine seconds ahead of teammate Carlos Reutemann and win his first Long Beach Grand Prix, as the 1981 season finally began after a winter of controversy and legal battles. It was the third consecutive Grand Prix win for Jones, and his second consecutive in the United States, after seizing the 1980 Drivers' title with season ending wins in Montreal, Canada and Watkins Glen, New York.

The off season had seen FISA (La Fédération Internationale du Sport Automobile) and FOCA (the Formula One Constructors' Association) in conflict, ostensibly over FISA's scheduled ban of aerodynamic skirts on the cars, but also over financial control of the sport. After threatening to institute their own championship, FOCA agreed to the skirt ban on assurance of their continued control of the sport's finances and FISA's commitment to a four year period of stability in the rules. Just 10 days prior to the season opening race in Long Beach, the Concorde Agreement was signed in Paris, allowing all of the teams to appear. In the meantime, the South African race, run in February under FOCA's pre agreement version of the rules, had been deprived of its World Championship status by the dispute, and the Argentine race, originally scheduled in January, was moved to April.

In addition to the new rules, Goodyear announced in December that it intended to withdraw immediately from all involvement in European racing. When the teams arrived in Long Beach for the first championship race of the season, the Friday morning practice sessions were filled with frantic activity. Larger wings, softer springs and revised sidepods were in evidence for nearly everyone, trying to make up for the absence of the banned skirts. With all teams also using Michelin tires as well, many drivers were struggling to come to grips with a totally new set of challenges.

The street circuit had been slightly modified from the year before; the second left hander on Pine Avenue had been made a single apex corner instead of a double apex.

On Saturday, yet another legal issue arose over the new twin chassis Lotus 88, designed by Colin Chapman and Martin Ogilvie. A protest was lodged by a majority of the teams, although they did not specify what rules it was breaking. The car was initially approved by the FISA technical staff and passed by the scrutineers, allowing it to take part in Friday practice. Ultimately, however, the teams' appeal was allowed, the car was banned from the rest of the weekend and Lotus had to qualify and race the more conventional Lotus .

On the track, in final qualifying, Ricardo Patrese and Alan Jones traded the top spot back and forth several times during the session. Patrese finally managed to take the pole, clinching his first ever and the first (and only) for his Arrows team, by .01 seconds.

Jones's Williams teammate Reutemann was third, followed by Nelson Piquet's Brabham, the Ferrari of Gilles Villeneuve and Mario Andretti in his first race for Alfa Romeo. The all American Tyrrell driver team had Eddie Cheever in eighth place, but Kevin Cogan missed the final qualifying spot by .07 seconds. It was the first time a Tyrrell had ever failed to make the grid.

GRAND PRIX: Patrese Gets the Pole

Continued from First Page

Patrese on the pole is a surprise even to him.

"It's a surprise for me," said an elated Patresse, who will be 27 next month. "I think it's a surprise for everybody."

It is the first time an Arrows has sat on a pole since Oliver introduced the car to Formula One racing four years ago. Its best finish was at Long Beach last year when Patrese chased Piquet across the line in second place.

Former Teammates Battle

"When they took away the skirts it seemed to hurt most of the cars but for some reason our car ran quite well," said Oliver, a former English racer who won the 1974 Can-Am championship. That was the same year a teenaged Petrese first attracted notice as the world karting champion.

"The best I had hoped for was to start beside Jones on the front row," Patrese said. "I am surprised to be ahead of him. Now I must concentrate on staying ahead of him. He is the world champion and a hard man to fight. I know, I have had many fights with him on the track."

Associated Press

Riccardo Patrese of Italy celebrates his pole-winning ride with Penthouse magazine "pet" Dominique Maure Saturday at Long Beach.

final lap of the qualifying session, earned the 24th and final berth.

Andretti, the 1978 world champion in a Lotus, will start his first race as an Alfa Romeo driver from the sixth position. Ahead of him are Patrese, Jones and three former Long Beach winners, Reutemann, Piquet and Villeneuve.

The on-again, off-again appearance of Colin Chapman's innovative Lotus 88 ended in mid-day when Sport Car Club of America stewards black-flagged it off the course during a practice session. The cars, with a secondary chassis and suspension system bonded to a primary chassis and another suspension system, had been accepted as an entry Wednesday, protested Thursday and ruled illegal Friday. Chapman appealed the ruling and stewards said the car could practice and attempt to qualify pending a hearing on Chapman's appeal. Later, however, chief steward John Bornholdt reversed his decision and barred it from the competition.

"I unfortunately made an error when stating the car, under appeal, could run," Bornholdt explained.

Three familiar faces from past Long Beach races, 1979 world

Sunday's weather was perfect, but the first lap was not. Villeneuve made a wild charge down the outside off the grid and briefly took the lead, but he left his braking for the Queens Hairpin far too late. As he went wide, Patrese and the Williams pair of Reutemann and Jones all went through. Villeneuve was able to gather it in and rejoin in fourth, but Andrea de Cesaris did not, as he ran his McLaren into the back of both Alain Prost and Héctor Rebaque approaching the hairpin. After being hit, Prost's Renault slid across the track and shoved the Brabham of Rebaque into the wall. Prost and de Cesaris were out on the spot, but Rebaque was able to continue after pitting for four new tires. After one lap, the order was Patrese, Reutemann, Jones, Villeneuve, Piquet, Didier Pironi, Cheever and Andretti.

Patrese's retirement with a spark box failure left Reutemann with a three second lead over teammate Jones, who immediately began closing the gap by half a second per lap. Any questions about team orders letting the number one driver through were soon answered. On lap 32, while lapping Marc Surer's Ensign, Reutemann slid wide in the esses on Pine Avenue, and Jones went through for the lead. Within 12 laps, the defending World Champion had stretched out a lead of ten seconds. At the same time, Reutemann was extending his lead over Piquet to 36 seconds.

Jones and Reutemann took the third consecutive one two for Williams. Jones collected $171,813 for the win while Reutemann's payday was $149,813. Piquet, having lost his shot at the leaders while bottled up behind Pironi, finished third, 35 seconds back. Mario Andretti thrilled the crowd with his fourth place, just ahead of compatriot Eddie Cheever's Tyrrell in fifth. It was the first time for two Americans to finish in the points since Andretti and Mark Donohue at the 1975 Swedish Grand Prix.

On another note, Clay Regazzoni, injured last year, has sued the Grand Prix Corporation and others for $20 million, citing negligence and bad track design. He later withdrew the suit, claiming he had been influenced by other and apologized to Chris Pook, joining him in a glass of champagne at Monaco years later.

Regazzoni Files Suit in Excess $20 Million

By RICH ROBERTS, *Times Staff Writer*

The lawyer for former Swiss race driver Clay Regazzoni, while acknowledging the risks of the sport, claims his client is due more than $20 million for the accident that left him a paraplegic a year ago.

William P.A. Camusi of Los Angeles charges the Grand Prix Association of Long Beach and others with "negligence" claiming an escape road was too short and was cluttered with two previously disabled cars, causing Regazzoni to crash into a tire and concrete barrier at a speed "in excess of 160 mph."

"In other words," Camusi says, "it's dangerous enough. You don't have to do things to make it more dangerous."

Camusi filed the suit in Los Angeles federal court Feb. 25. The next Grand Prix of Long Beach is scheduled Sunday, March 15. Camusi denied the timing was more than coincidence.

Patient And Relaxed Alan Jones Wins Long Beach Grand Prix

LONG BEACH, Calif. (UPI) — Listening to Alan Jones, you'd think he had just taken his kids for a Sunday drive along the ocean.

The 34-year-old Australian talked of being relaxed, of being patient. Next time you're driving through the downtown area of a city at speeds up to 190 mph, try relaxing and being patient.

But Jones said those were the prime ingredients in his victory Sunday in the Long Beach Grand Prix, the first race of the 1981 Formula One World Championship.

Jones' margin of victory over teammate Carlos Reutemann of Argentina was 9.19 seconds, well off the winning margin of 45 seconds by Nelson Piquet in last year's Long Beach Grand Prix, but still an enormous margin by Formula One and Grand Prix standards.

Jones started in second place behind Riccardo Patrese of Italy but fell to third on the first lap as Reutemann bolted past him. On the 26th lap, Patrese's car began to falter and Reutemann and Jones moved into the 1-2 positions.

"I had a dreadful start," said the 34-year-old Jones, "so I figured I'd sit back and relax and be patient for about 20 laps, and then see what was in front of me.

"Carlos began having some problems in the S-turns swinging out a little wide, and he had to come off the accelerator. I saw the opening and I took advantage of it."

The triumph got Jones' William-Ford team off to a great start for the 1981 season and Jones knows he's going to be the driver to beat. But he also knows the 1981 World Championship won't be handed to him.

"I don't know if I can repeat as champion," he said, "but I know I'm going to give it a bloody good try."

Reutemann, who won the race in 1978, finished 26 seconds ahead of third-place finisher Piquet. Americans Mario Andretti and Eddie Cheever were fourth and fifth.

Riccardo Patrese takes his car on the wide track around the first turn on lap one of Sunday's Toyota Grand Prix of Long Beach. Patrese, who started in the pole position, left the race in the 27th lap, and Australia's Alan Jones came on to win. (UPI)

(See all the race results on the attached DVD)

JAYNE KAMIN / Los Angeles Times

A smiling Alan Jones holds aloft trophy after driving to easy victory in Long Beach Grand Prix.

News-Pilot
Fri., March 6, 1981 C

Formula One is the biggest, but not the only part of the Grand Prix weekend in Long Beach. Sidecars also will be performing March 15.

Agility, timing crucial in motorcyle sidecar racing

Sports Illustrated

THAT JONES BOY STRIKES AGAIN

Read the Grand Prix International, Autoweek, Sports Illustrated, Autocar and Motor Sport articles on the attached DVD

Watch the video on our you tube channel
www.youtube.com/@racinghistoryproject2607

Mario Andretti

Alan Jones

Eddie Cheever

Elio de Angelis

Jacques Lafitte

Nigel Mansell

Hector Rebaque

Siegfried Stohrr

Nigel Mansell

Alain Prost

Didier Pironi

Gilles Villeneuve

1982

The Toyota Grand Prix of Long Beach, race three of the sixteen race World Championship, was held on April 4, 1982.

The previous Saturday had the 10K charity run on the circuit followed by the Concour d'Elegance in the paddock and a BMX bicycle motocross at the intersection of Linden and Shaw. Sunday was the Chili Cook Off in the paddock at 10:00 AM followed by another BMX bicycle motocross from 10:00 AM to 4:00 PM. The annual art show was held at Harbor Bank on Tuesday, then the race weekend began on Thursday with the Formula One Concours and Race Car Show from 11:30 AM to 4:00 PM on Pine Avenue and Prix View event in the Formula One garage at 5:30 PM. The Formula One garage opened on Friday morning as did the Grand Prix Expo. The day consisted of practice and qualifying sessions for all groups.

The Toyota Celebrity Race, at 11;45 AM on Saturday was won by Bruce Jenner followed by Dan Gurney and Josele Garza. The Bell Helmets BMX Exhibition took place at 12:20 PM followed by the Nike 10K run at 2:10 PM and the Formula Atlantic Race at 4:30 PM, won by Geoff Brabham with Roberto Moreno in second and Al Unser Jr. third. Sunday had a Malibu Grand Prix Exhibition at 9:15 AM, the 250cc motorcycle race at 10:45 AM and the Bridgestone Pro Kart Race at 11;20 AM with the Formula One Race starting at 1:00 PM.

In his third race since returning from a self imposed two year retirement, a crowd of 82,000 watched Austrian Niki Lauda win, ahead of Keke Rosberg. It was the 18th victory of Lauda's career, and his first for McLaren. Canada's Gilles Villeneuve crossed the line in third, but he was disqualified after the race when a protest of his Ferrari's rear wing was upheld by the officials. Lauda's win paid $178,675.

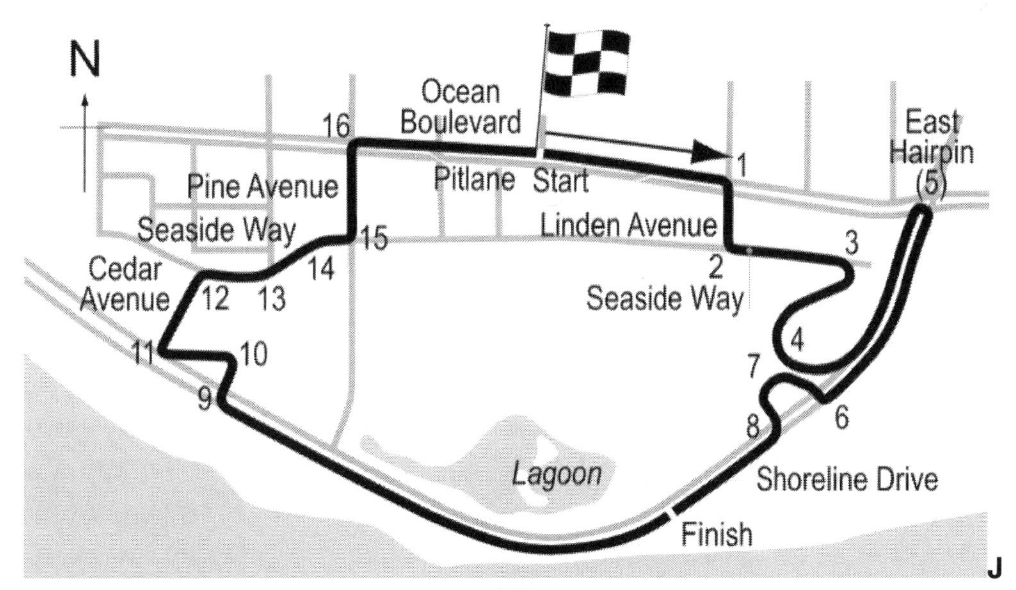

Just five days after the previous race in Brazil, Carlos Reutemann had shocked his boss Frank Williams, and everyone else in the paddock, by announcing his retirement. When former World Champion (for Williams) Alan Jones insisted he was not available, Williams contacted another former Champion, American Mario Andretti. His commitment to the Patrick Indy racing team posed no conflicts, so he agreed to drive the second Williams for the weekend, saying, *"I had nothing else to do, so I accepted."*

Significant changes had been made to the course since the previous year's race. The Queen's Hairpin at the end of Shoreline Drive had been transformed into a right-angle turn that led into a new section of track with several demanding corners, leading up to Ocean Boulevard. On the other end of the course, the short straight at the bottom of the hill from Linden Avenue had been lengthened and a chicane had been inserted near the beginning of the curving Shoreline Drive straight, in anticipation of the pits being moved there from Ocean Boulevard. The changes increased the length of the lap only slightly, but added about ten seconds to the previous year's times as the drivers became acclimated to the new layout.

Rosberg races to the pole position at LBGP — for 24 hours, at least

By JIM SCHULTE
Sun Staff Writer

LONG BEACH — After what he accomplished Friday afternoon during qualifying for Sunday's Long Beach Grand Prix, it's now OK to call Keke Rosberg the "Flying Finn" — for exactly 24 hours.

At the end of that period today — when the little hand is on the 2 and the big hand on the 12 — Rosberg will have had to duplicate both his feats of Friday in order to keep his moniker.

What he did yesterday was to push his Williams SW07 to a speed of 86.750 mph to qualify as the polesitter for this seventh edition of the LBGP. That speed also set a course record. Unfortunately for Rosberg, both those items have an asterisk beside them.

You see, Friday's session was only the first of two qualifying trials, with the second beginning at 1 p.m. today. And as for the lap record, that comes about by virtue of the fact that this is a new course, redesigned and lengthened to 2.13 miles from last year's 2.02

who coaxed a speed of 85.263 from his Brabham B149D. Next was Alain Prost, this year's winner in Brazil, who turned 85.262 in his Turbo Renault.

Then came 1979 LBGP winner Gilles Villeneuve in a Turbo Ferrari (85.248 mph), Nigel Mansell in a Lotus (85.089), Didier Pironi in the other Ferrari (85.082), Riccardo Patrese, last year's pole sitter, at 84.935 in his Grabham, Elio de Angelis in the second Lotus (84.668) and Bruno Giacomelli in an Alfa Romeo (84.571) rounding out the top ten.

The fact that the turbo-powered machines didn't do as well as they normally do here surprised many, including Rosberg.

"The turbos are a different ballgame altogether," he said. "I was behind Prost today and as soon as he hits fourth gear he's gone. It's unbelievable. You think that your car is actually racing fairly fast. And then you see how the other one (turbo) just disappears in front of you. It's really quite impressive.

"But it evens out. The turbos will be difficult to drive here with all that power."

In Saturday's qualifying, the cars running on Michelin tires had a decided advantage over the Goodyear teams, though the Michelin men had all learned from Lauda's times on Friday that their harder race tires were faster than the qualifiers. Seeing this development, Lauda intentionally used only one set and kept a brand new set for Sunday's race. Lauda topped the charts through almost the entire session, but, after crashing into a wall early on, Andrea de Cesaris beat Lauda's time by .012 of a second, three minutes before the session ended. It was the Italian's first pole position (his *only* pole in 208 career starts), and he was ecstatic. At the time, de Cesaris was the youngest driver to achieve pole position, a record beaten by Rubens Barrichello at the 1994 Belgian Grand Prix.

Italian grabs pole at Long Beach Grand Prix

LONG BEACH (AP) — Italy's Andrea de Cesaris, 22, the youngest driver in Formula One racing this year, drove an Alfa Romeo to the pole position Saturday for the Toyota Grand Prix of Long Beach.

De Cesaris, who never has held a pole position or won a Grand Prix in 18 previous races, trailed two-time world champion Niki Lauda of Austria and Frenchmen Rene Arnoux and Alain Prost until his last lap around the twisting 2.13-mile circuit through the streets of downtown Long Beach.

His fast lap of 87.819 mph knocked Lauda into the second-spot on the 26-car grid for today's race.

The dark, handsome De Cesaris burst into tears moments after he came into the pits following the checkered flag at the end of the qualifying session.

"I was calm when I came in," de Cesaris said. "But, when I saw all my mechanics and the other people on the team running toward me so excited, I started crying. It was wonderful."

He said he knew the last lap was faster than he had turned before because "I wasn't overtaking anybody. I had a clear track and I just felt it was going to be real quick."

Lauda, who is in his third race since coming out of a two-year retirement, piloted a McLaren to a fast lap of 87.696 midway through the last of two hours of qualifying.

Renault teammates Arnoux and Prost, both driving turbocharged cars that have in the past been at a disadvantage on a tight, twisting course like Long Beach, wound up third and fourth.

holds virtually every race and qualifying record at the old distance of 2.02 miles — crashed with about 20 minutes left in Saturday's time trials. He was not injured, but his Brabham car suffered extensive damage to the right front side and may not be driveable Sunday.

Since the Brabham team did not bring a backup car, Piquet may wind up starting the race in the car of teammate Riccardo Patrese of Italy, who qualified his car 19th.

Keke Rosberg of Finland, who put his Williams racer on the provisional pole in Friday's qualifying session, had handling problems Saturday and was unable to improve on his previous high of 86.578. That was still good enough to keep him in the eight position, next to Canadian Gilles Villeneuve, who put his turbocharged Ferrari on the inside of the fourth row with a speed of 86.667 mph.

Mario Andretti, coming out of a brief Formula One retirement to run at least this one race for the Williams team, managed to move up to the 14th spot on the grid. The American driver, a former world champion, had a fast lap of 85.706.

The only other American in the field, Eddie Cheever, driving a Talbot-Ligier, wound up just ahead of Andretti in 13th at 85.833 mph.

Other than an Alfa winning the pole, the biggest surprise was the performance of Renault's turbocharged cars. Although the turbo engines are much faster than conventional, normally aspirated engines on tracks with long straightaways and sweeping turns, a circuit like the short, 13-turn Long Beach course usually poses problems.

found additional speed Saturday by eliminating some extra filters in the turbocharger system.

That increased the flow of air through the turbocharger and reduced the lag time.

"Our fastest laps were on racing tires (rather than qualifying tires) and we were very fast in the morning (practice session) with full fuel loads. That combination makes us very confident," said Arnoux.

"It's very difficult to use the turbo here. It seems like we are always either in second gear or using the brakes. But I'm very happy to be starting in the second row."

DeCesaris, whose new red and white Alfa Romeo 182 is powered by a conventional V8, said he likes the idea of starting in front.

"I've never done it before, but there should be less trouble there," he explained. "The car I think will be reliable. I expect to finish the race."

The race, which starts in the middle of the back straightaway and ends at the finish line in front of the pits on the front straightaway, will run 75.5 laps or two hours, whichever comes first.

This year's qualifying times were quite a bit slower than last year because of changes to the course forced by construction of a new hotel within the circuit. The track was lengthened, but also includes five new turns and a short chicane — kink — which shortened the longest straightaway by about one-third.

* * * *

LONG BEACH (AP) — Geoff Brabham easily raced to a victory in the North American Formu-

As the cars formed on the grid for the start, Lotus driver Elio de Angelis lined up on the wrong side, claiming he was waved into the wrong place. He quickly backed out of the spot, bumping his teammate Nigel Mansell behind him. When Mansell put his car into reverse, thinking that de Angelis was coming back further, the green light came on. As a result, Mansell claims to be the only driver to have started a race in reverse. Everyone got away cleanly, though Mansell found himself near the back of the field. At the front, de Cesaris made an excellent start, jumping into the lead ahead of Lauda and René Arnoux.

Niki Lauda of Austria sweeps past Bruno Giacomelli during yesterday's Long Beach Grand Prix auto race. Lauda eventually won the race. (AP Photo)

At the end of the first lap, de Cesaris led by two seconds, followed by Arnoux and Lauda, On lap six, with the Italian beginning to stretch his lead slightly, his Alfa Romeo teammate Giacomelli closed up on Lauda, who was right behind Arnoux. As the three cars approached the hairpin, Giacomelli made a run down the outside of Lauda, locked up his brakes and slid into the back of Arnoux's Renault. Both cars were out, and de Cesaris now led Lauda by 5.7 seconds, with Villeneuve in third. Prost hit the inside wall while under braking for the right hander at the end of Ocean Boulevard and was immediately out.

Lauda began to cut into de Cesaris' lead, setting the race's fastest lap in the process. On lap 15, de Cesaris was held up by Raul Boesel's March in the chicane entering Shoreline Drive as he came up to lap him. This gave Lauda the momentum he needed to sweep by into the lead at the end of the straight. Andretti lost it in the marbles of tire rubber that were collecting off line and damaged his suspension against the wall in turn four. De Cesaris distracted by smoke from an engine fire, flew off the road into the turn five wall on lap 34, ripping off two wheels and the right sidepod. This left Lauda almost a full minute ahead of Rosberg, with only Villeneuve, Alboreto and Cheever also on the lead lap. Lauda came home nearly 15 seconds ahead for his second win in the United States, along with the 1975 Watkins Glen race, with Rosberg second.

(See all the race results on the attached DVD)

Lauda outlasts Rosberg in Long Beach Grand Prix

LONG BEACH, Calif. (AP) — Cautiously protecting a big lead through the last 10 laps of the Toyota Long Beach Grand Prix, Niki Lauda of Austria bounced and skidded to his first Formula One victory in nearly four years.

Lauda, 33, driving a McLaren racer for only the third time since his return from a two-year retirement, won the 18th Grand Prix race of his career.

He easily outlasted new Grand Prix point leader Keke Rosberg of Finland and the Williams team in the finishing laps on the 2.13-mile course through the streets of Long Beach.

There were complaints by several drivers, including Lauda and Rosberg, about the course, which was reshaped from previous races because of the construction of a hotel within the circuit. Some newly-laid sections of asphalt broke up as the 75.5-lap race progressed.

That caused both leaders to slow down by 15 to 20 mph compared to speeds in the early part of the nearly two-hour endurance test for the exotic cars. And only 11 of the 26 starters were running at the finish.

"I slowed down because it was important to win," said Lauda, who last picked up a victory in the 1978 Italian Grand Prix. "The (time) difference didn't mean a thing," he added, speaking of his final 14.660-second margin over Rosberg.

"There was no racing for the last 20 laps. It was everybody just trying to save themselves amd finish the race," said Rosberg, who took over the point lead after three rounds of the Formula One series with 14 points. That's one more than Alain Prost of France, who lasted just 10 laps in his turbocharged Renault, and two points ahead of Lauda.

"If you were five centimeters off the lipe, you crashed," Lauda said of the asphalt torn up by the cars on new sections of the revamped course. "I think it is not a good idea to resurface the course just before a race."

Lauda averaged 81.4 mph, far below the record of 88.47 mph set by Brazilian Nelson Piquet of Brabham in 1980 on the old course. Lauda kept turning laps in 1 minute, 35 seconds or less most of the race, then slowed to laps of 1:50.2 to 1:58.3 in the final six trips around the course.

Rosberg, however, was unable to do much better. He kept creeping closer, but never really made a serious challenge.

Gilles Villeneuve of Canada was third in a turbocharged Ferrari, more than one minute behind the winner. Riccardo Patrese of Italy was fourth in a Brabham, and Michele Alboreto of Italy was fifth driving a Tyrell.

Only 22-year-old Italian Andrea de Cesaris and his Alfa Romeo had been able to match Lauda for the first half of the race before crashing his car on the 34th lap.

Nation's best cyclists invited to $5,000 race

Forty of the country's best motorcyclists will be hitting speeds of up to 140 m.p.h. April 4 in the 250cc motorcycle race, a new addition to the Toyota Grand Prix of Long Beach weekend.

The 250cc class has been a proving ground for such famous motorcycle greats as three-time 500cc World Champion Kenny Roberts, 1981 500cc World runner-up Randy Mamola and AMA Champions like Gary Nixon and Freddie Spencer, new star of Team Honda.

The invitational event, organized by AFM Pro Promotions, offers a purse of $5,000, with the winner banking $1,250. And when the checkered flag falls, the winner may come from right here "at home."

The field at Long Beach will feature such Southern California stars as David Emde, the past national champion from Oceanside; Gill Martin, San Diego, and Harry Klinzmann from Anaheim Hills. Northern California motorcycle stars include Fred Winter, Fred Merkel and John Williams.

The distaff side is led by Christine Baur, a stunt driver from Topanga Canyon. Also entered is John Glover, who recently finished sixth at Daytona Beach, and John Long from Florida, who finished fourth in a 750cc race at Long Beach in 1977.

The bikes they'll pilot are sophisticated creations designed for only one thing — high-speed racing. Costing approximately $7,000 each, they weigh 240 pounds and can top 155 m.p.h. Nearly all the entrants will be powered by 60-horsepower, 15.2-cubic-inch Yamaha engines.

Because of the minmal horsepower difference, the Long Beach race will feature plenty of tightly packed racing with lots of passing and according to one 250cc expert, a little NASCAR-style bumping on the turns.

The Long Beach race may well be won on the circuit 12 challenging turns. To get a fast lap, the 250cc riders will use their knees as "rudders" actually touching the pavement to maintain a balance point with the ground. And, they'll be dragging those knees around turns at speeds up to 130 m.p.h.

The 250cc class is sanctioned nationally by the American Federation of Motorcyclists in Buena Park. Although the Long Beach race, an invitational event, will not count toward the national championship, eight national events around the nation will determine the AFMs 250cc Champion on the basis of races run and points scored.

Last year's 250cc champion was Eddie Lawson, now the leading international Kawasaki rider.

The AFM also sanctions six-hour endurance races with a wide spread of machines entered, from Formula One monster bikes to stock single-cylinder street bikes. About two dozen endurance races are held every year in the U.S. including, in 1981, the Budweiser Six-Hour at Riverside International Raceway.

Practice for the 250cc motorcycles will be at 11:40 a.m. April 2 and 3:30 p.m. April 3. The race will get under way at 10:45 a.m. April 4.

Football stars head Pro/Celeb race

Sure, Joe Montana and Vince Ferragamo are good quarterbacks, but can they handle a race car?

The answer will come Satur-

rounds out the pro-football contingent for the race.

The three football stars will be chased by three of the great names in American racing. Dan Gurney, who still holds the

CBS-TV race commentator takes an early lead in the 1981 Toyota Pro/Celebrity race. Film star Bob Hays went on to win the event and returns this year to defend his title.

$10,000 purse for kart racers

art racers roar over the Long Beach Grand Prix course in action from last year. This year's drivers will compete for a $10,000 purse.

Watch the video on our you tube channel - www.youtube.com/@racinghistoryproject2607 and read, the Grunion, Sports Illustrated and Motor Sport articles on the attached DVD

Bruno Giacomelli

Mario Andretti

Elio de Angelis

Andrea de Ceasaris

Nigel Mansell

Keke Rosberg

Nelson Piquet

Gilles Villeneuve

1983

The Toyota Grand Prix of Long Beach, round two of the Formula One World Championship, was held on March 27, 1983. This would be the last Formula One race at Long Beach, with the racing shifting to CART in 1984. Again there was a slight circuit revision, freeing up Ocean Boulevard and moving the pits to Shoreline Drive.

The race week began with the Rusty Pelican 10K run at the Queen Mary and then a Concours d'Elegance and bicycle motocross. Tuesday featured an art show at Harbor Bank and the Hyatt Regency Grand Prix Celebrity Golf Tournament ar Recreation Park.

Then on Thursday there was the F-1 Concours and Race Car Show on Pine Avenue, followed by the Prix View Round Up in the Formula One garage. The Grand Prix Expo opened on Friday along with practice and qualifying for all races; Formula One, 250cc motorcycles, Toyota Celebrities, Bridgestone Pro Karts and Super Vee.

Saturday was the Toyota Celebrity Race at 11:40 AM with the Super Vee Race at 4:00 PM.

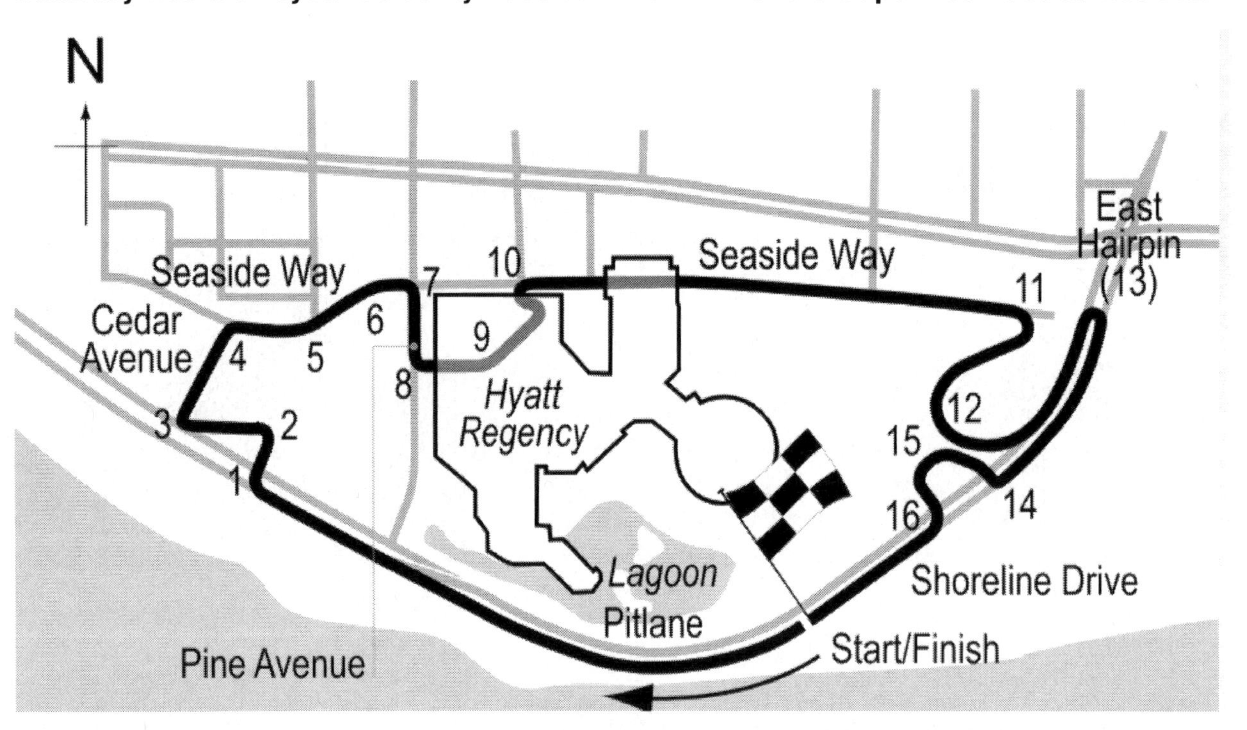

Sunday, race day, had the Bridgestone Pro Kart Race at 11:20 and Formula One Race at 1:30 PM. And after the event, there was the Tecate Grand Prix Western Festival at Shoreline Village with Chili and Ribs Cook Offs.

Dan Gurney won the Toyota Celebrity Race, followed by Parnelli Jones and Gene Hackman. Michael Andretti won the Super Vee Race, followed by Hubert Phipps and Jerrill Rice.

John Watson, the most experienced driver in Formula One, put on a brilliant drive from 22nd starting position to win the Grand Prix in one of the most stunning performances in Formula One racing history. He drove a John Barnard designed McLaren MP4/1 C Cosworth to a 27 second cushion over Marlboro McLaren teammate Niki Lauda of Austria, the defending champion under warm and sunny conditions. As practice began on Friday, two bumps where the circuit rejoined the old layout at the end of the Seaside Way straight were causing problems. Some teams were concerned that the suspension on their cars would not last more than a few laps under race conditions. René Arnoux (Ferrari) was the first to go over the bumps flat out and his 1:26.935 led Alain Prost (Renault), Patrick Tambay (Ferrari) and Riccardo Patrese (Brabham) on the day's timing chart, while Nelson Piquet (Brabham), Lauda and Watson found their Michelin qualifying tires virtually useless and set poor times.

Overnight repair work smoothed the problematic bumps. Tambay grabbed his first pole with a lap of 1:26.117, the only lap to beat teammate Arnoux's Friday time; Keke Rosberg (Williams) took third with 1:27.145.

At the start, Tambay held the lead at the first corner. Rosberg, immediately behind him, tried to squeeze through the middle of the all Ferrari front row. He touched Arnoux's right front with his left rear as he swung wide, but both continued, with Rosberg in second, followed by Laffite and Arnoux. Rosberg spun later in the lap while attempting to overtake, but continued without damage.

Laffite was now in the lead, with Patrese in second with the McLarens lying third and fourth. Watson got by Lauda at the end of Shoreline Drive and was 20 seconds behind the two leaders. With Watson closing the gap to the front and Laffite's tires going off quickly,

Patrese challenged Laffite for the lead. He slid wide, and Watson and Lauda both passed before he rejoined the track. Soon after, the McLarens passed Laffite as well, and, from 22nd and 23rd on the grid, were now first and second.

At the hairpin, the Williams and Ferrari swapped places around Cheever, as Arnoux went from sixth to fourth in one corner. On the next lap, however, Cheever lost fifth place when he pulled off with a broken gearbox. With just three laps to go, Patrese retired from third place when his distributor broke.

Lauda, suffering from a worsening cramp in his right leg, could not challenge Watson in the later stages, and the Ulsterman came home nearly half a minute ahead for his fifth victory. It was the farthest back from which a modern Grand Prix driver had ever come to win a race, and it remains the only time in F1 history a driver has won who qualified 20th or worst.

In Long Beach Grand Prix

Frenchmen sweep front row

LONG BEACH (UPI) — Ferraris driven by Frenchmen Patrick Tambay and Rene Arnoux swept the all-important front row starting positions Saturday for today's Long Beach Grand Prix.

Tambay, who won the 1982 German Grand Prix for his only Formula One victory, took the pole

time of 1:26.936 (84.270 mph) but hung on to the No. 2 starting spot.

The Ferrari front row sweep marked the first time since 1978 the team qualified 1-2 for a Formula One race.

Another pair of teammates, defending world champion Keke Rosberg of Finland and Jacques

1:27.818, were both 1½ seconds slower than the Ferraris. The Williams cars are both powered by normally aspirated Cosworth engines while the Ferraris have turbocharged power plants.

Irishman makes history in GP

LONG BEACH (UPI) — John Watson of Ireland scored a longshot victory in the Long Beach Grand Prix, driving his McLaren from an almost hopeless 22nd starting position to a 27-second win over teammate Niki Lauda of Austria.

Watson's victory Sunday, in 1:53.34.9 at an average speed of 80.62 mph over the 157-mile city street course, marked the first time in Long Beach Grand Prix history a winner came from other than a front-row starting position.

Watson took the lead on the 45th lap after trailing the early leaders by as much as 22 seconds and was never seriously threatened.

Starting 22nd and winning is a Grand Prix first. The previous record also belonged to Watson, who won the 1982 Detroit Grand Prix after starting 17th.

Rene Arnoux of France was third in a Ferrari while countrymate Jacques Laffite finished fourth in a Williams, ahead of Marc Surer of Switzerland, in an Arrow, and a Theodore driven by Johnny Cecotto of Venezuela.

bay pushed his Ferrari to a six-second lead over defending World Champion Keke Rosberg of Finland until the 25th lap.

There, Rosberg attempted to overtake the leader on a hairpin turn. As Rosberg moved inside, his Williams clipped Tambay's Ferarri, nearly flipping it.

"Rosberg was running much too hot chasing me, and he lost his cool in the turn and ran into me," claimed a fuming Tambay.

Rosberg then drove around Tambay's careening car but in the next turn Rosberg collided with Jean-Pierre Jarier of France and both machines spun off the course and out of the race.

Laffite then inherited the lead ahead of Riccardo Patrese of Italy while Lauda moved up to third, followed by Watson, who was more than 22 seconds behind the leader. Watson passed Lauda on the 32nd lap and cut Laffite's lead from 19 seconds to nine seconds, then to four seconds and finally, on the 45th lap, he surged ahead.

Watson creditied the win to his selection of tires.

work well here, so we took the gamble and it paid off."

Both McLarens experienced tire problems in qualifying which resulted in their starting far back in the 26-car field.

"Our problems were basically tire temperatures," Watson said. "We weren't getting the heat we needed to get a good grip. But our race tires worked fine. Niki chose just a little different tire compound than I did."

The victory was Watson's fifth on the Formula One circuit. His last victory came in last year's Detroit Grand Prix.

One of the best performances of the day was turned in by the United States' Eddie Cheever, who overcame two pit stops with his turbo Renault to finish fourth.

The only other American in the race, Danny Sullivan, finished eighth in a Tyrrell.

A crowd of about 75,000 watched the seventh and what may have been the final Long Beach Grand Prix for Formula One cars. Race officials scheduled a news conference today and were expected to an-

(View all the race results on the attached DVD)

Moves up from 22nd position

Watson captures Grand Prix win

LONG BEACH, Calif. (UPI) — John Watson of Ireland scored a longshot victory in the Long Beach Grand Prix, driving his McLaren from an almost hopeless 22nd starting position to a 27-second win over teammate Niki Lauda of Austria.

Watson's victory Sunday, in 1:53.34.9 at an average speed of 80.62 mph over the 157-mile city street course, marked the first time in Long Beach Grand Prix history a winner came from other than a front-row starting position.

Watson took the lead on the 45th lap after trailing the early leaders by as much as 22 seconds and was never seriously threatened.

Starting 22nd and winning is a Grand Prix first. The previous record also belonged to Watson, who won the 1982 Detroit Grand Prix after starting 17th.

Rene Arnoux of France was third in a Ferrari while countrymate Jacques Laffite finished fourth in a Williams, ahead of Marc Surer of Switzerland, in an Arrow, and a Theodore driven by Johnny Cecotto of Venezuela.

Watson, at 36 the second oldest driver in the race, was one of three leaders during the 75-lap race. Pole-winner Patrick Tambay pushed his Ferrari to a six-second lead over defending World Champion Keke Rosberg of Finland until the 25th lap.

There, Rosberg attempted to overtake the leader on a hairpin

Sports scorecard
— Page 6

turn. As Rosberg moved inside, his Williams clipped Tambay's Ferarri, nearly flipping it.

"Rosberg was running much too hot chasing me, and he lost his cool in the turn and ran into me," claimed a fuming Tambay.

Rosberg then drove around Tambay's careening car but in the next turn Rosberg collided with Jean-Pierre Jarier of France and both machines spun off the course and out of the race.

Laffite then inherited the lead ahead of Riccardo Patrese of Italy while Lauda moved up to third, followed by Watson, who was more than 22 seconds behind the leader. Watson passed Lauda on the 32nd lap and cut Laffite's lead from 19 seconds to nine seconds, then to four seconds and finally, on the 45th lap, he surged ahead.

The victory was Watson's fifth on the Formula One circuit. His last

UPI photo

TAKING THE CHECKERED: John Watson of Ireland and his Marlboro McLaren International cross under the checkered flag at Sunday's eight annual Toyota-Long Beach Grand Prix. Watson came up from 22nd position

Long Beach Grand Prix may be seeing checkered flag

LONG BEACH (AP) — Rising costs and declining attendance apparently are threatening the Long Beach Grand Prix, causing some to forecast that the race on local streets is about to go under and be replaced by the standard oval track.

The Grand Prix, which starts Friday, stands to lose not only the international character of its race, but possibly its main sponsor — Toyota Motor Sales — if the move is made. But insiders say the prospects are dim that the change can be avoided.

"CART (Championship Auto Racing Teams) is all set to replace the Formula One entry at Long Beach in '84, the announcement supposedly to be made within 24 hours of the finish of this year's Grand Prix," according to the January issue of National Speed Sport News.

The president of the Long Beach Grand Prix Association, Chris Pook, refused to be interviewed on the subject for a recent article in the Long Beach Press-Telegram.

The Grand Prix Association told the city council last July that this year's race will cost $3.65 million, or $350,000 more than last year. The increase is mainly in prize money and transportation fees for international driving teams, which alone tallies $1.75 million.

> 'The contract specifies that Toyota will sponsor a Formula One race. I can tell you this, one of the things that appealed to us in the first place about the Grand Prix is the international flavor. If Formula One is gone, that international flavor is gone.'
>
> — **Toyota spokesman**

Attendance for last year's race dropped 7,000 from the 1981 high of 82,000, although paid attendance was 180,000 for the entire three-day race, about 5,000 fewer than the year before.

It was the first decline in attendance since the race started in 1976.

The association says the decline is a result of the recession and hikes in ticket costs. General admission tickets are $20 this year, $4 more than last year. Three day reserve seats are selling for $65-$100, some $35 more than last year.

The Long Beach Grand Prix Association also lost $194,000 on its abortive attempt to televise the 1982 Grand Prix season in hour-long segments during the summer, according to a city auditor's report. The telecasts stopped for apparent lack of interest after two programs were distributed.

Meanwhile, Toyota Motor Sales has declined to state how much it is putting into the race, but officials say the international character of a Formula One competition was a large draw to support the Long Beach contest.

"The (sponsorship) contract specifies that Toyota will sponsor a Formula One race," said Toyota spokesman Art Garner. "I can tell you this, one of the things that appealed to us in the first place about the Grand Prix is the international flavor. If Formula One is gone, that international flavor is gone."

The company has made no comment on what it plans if the city switches to an oval race.

"The only thing we can do is wait and see. We'll sit down with Chris (Pook) and discuss it on the 28th of March," said Lee Sawyeer, auto sports manager for the company.

City officials have noted that CART racing is less expensive because travel costs do not include shipping cars, materials and drivers from foreign countries.

But the city would then have to compete with nearby Riverside, where the Riverside International Raceway sponsors the Los Angeles Times California 500 race every year.

Nugent Tops Long Beach Hit Parade

By RICH ROBERTS, *Times Staff Writer*

If your car was anywhere near Long Beach Saturday, check for dents. Ted Nugent probably hit it.

The Detroit-born rock star is known as the "Motor City Madman," a suitable pseudonym when he drives in the Toyota Pro/Celebrity Race the day before the Grand Prix.

Nugent won the celebrity division Saturday, finishing fourth overall behind professionals Dan Gurney, Parnelli Jones and Gene Hackman, who somehow stayed out of his way to finish in that order.

"I had absolutely no game plan," Nugent said. "I just went out there with a chance to wreck everybody else's car."

He succeeded so well that halfway through the race the clerk of the course gave him the black flag. Normally, that means a driver's car is leaking oil or is otherwise endangering rivals. In Nugent's case, the driver was the menace.

"Did you see how many contacts he had?" a race official asked, incredulous.

The official said the flag was only a "warning" for reckless driving—"I doubt if he even knew what we were telling him"—and in any case Nugent ignored it, continuing merrily on his path of destruction.

"The streets of Detroit is where I learned my finesse," he explained to the press afterward.

Nugent's favorite target was actor Robert Hays, renewing their slam-bang rivalry of a year ago when Nugent took Hays into a wall and out of the race. He did it again Saturday—twice.

The first time was a classic racing move. Nugent overtook Hays on the third lap, locked him up broadside and rode him into leader Sam Moses entering a turn. As those two spun out, Nugent slipped past into the lead.

"It was self-bleeping-defense," Nugent said. "I haven't done anything I haven't seen the pros do."

Later, Hays caught up and passed as Nugent poked his nose into his door but was taken out again when Nugent rear-ended him.

No hard feelings, apparently.

"Ted drove real hard," Hays said, pausing to congratulate the winner. "He was something like he was last year—a wild man."

"My reputation precedes me," Nugent said.

The only quarter Nugent gave was to let pros Gurney, Jones and Hackman pass him from their starting position 30 seconds behind the amateur field. It was open season on everyone else.

Two-time world speedway motorcycle champion including a wild one when Jones brushed him at high speed on a straightaway, leaving Penhall pointing the wrong way.

"I almost tipped over," Penhall said.

"I gave him a thrill, I'll tell ya," Jones said gleefully.

Defending champion Bruce Jenner was a victim of the event's only woman driver—professional Margie Smith-Haas.

"I had the lead at the start but she T-boned me at the first turn," Jenner said, showing a huge dent in his right door. "I caught her and then she took me out a second time."

Hardly any car escaped the race unscathed. Remarkably, one of the cleanest was that of boxer Alexis Arguello, who was driving his first race.

"One time I got hit and spun three times," Arguello said. "It was great."

Nugent: "Alexis had a good idea. He thought before the next race we ought to have a pro-celebrity boxing tournament."

whether the strategy would
its narrower and tighter circ

"I know Ferrari will star
the conventional way," Tar
start light, perhaps they can
but it will be more difficult h
nearly a minute ahead on
backmarkers (cars in the
don't know if it could be don

No car has ever won at L

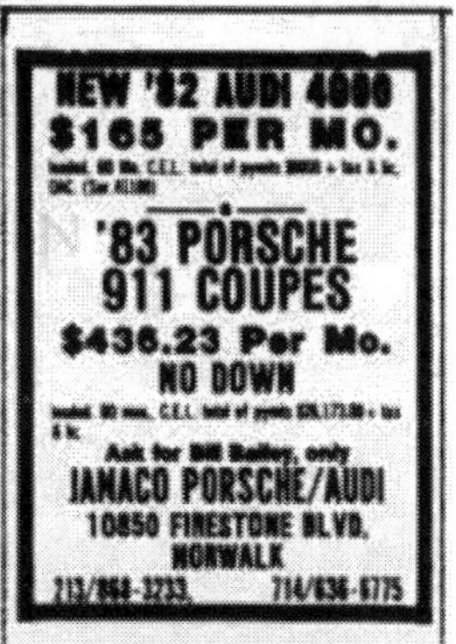

Watch the video on our you tube channel - www.youtube.com/@racinghistoryproject2607 and read Grand Prix International and Motor Sport magazine on the attached DVD

Raul Boesel

Ricardo Patrese

Keke Rosberg

Rene Arnoux

Corrado Fabi

Danny Sullivan

Michele Alboreto

Eddie Cheever

1984

The Toyota Grand Prix of Long Beach, held on April 1, 1984, was the opening race of the Championship Auto Racing Team (CART) season and the first time these cars had raced at Long Beach. Admission fees had been reduced by 22% due to the lower costs of bring Indy type cars here with general admission now only $15.

On the previous Saturday, the Concours d'Elegance was held at Shoreline Park. Sunday at 9:00 AM, the Tecate Chili Cookoff, at Shoreline Aquatic Park, started with a Chili Parade led by the Long Beach Junior Concert Band with cooking beginning at 1:00 PM. Their Ugly Dog Contest began at 1:00 PM.

Tuesday was the Grand Prix Charity Golf Tournament, matching local amateurs with celebrities and drivers, at Recreation Park. Friday was practice and qualifying for all groups. The Toyota Celebrity Race was at 11:45 AM on Saturday followed by the Russell Pro Series Race .

Sunday, the OK run started at the Queen Mary., then came the Bridgestone Pro Kart Race and the Bosch Super Vee Race and the main event, the CART race. NBC televised the event on a one week taped delay with reporters Johnny Rutherford and Bruce Jenner and announcer Paul Page..

Andretti Captures Pole in New Lola

He Beats Challenges of Younger Drivers, Including His Son

By SHAV GLICK, *Times Staff Writer*

Some things in auto racing never change.

Mario Andretti, who won the pole for the first-ever race through the streets of Long Beach nine years ago,

The right side was extensively damaged, but the Bignotti-Cotter team has a second car that will be a

Mario Andretti, who won the 1977 race when it was Formula One, won the pole position, took the lead at the start and led all 112 laps en route to a dominating victory, collecting $71,350.. The only other driver to finish the race on the lead lap was Geoff Brabham, who finished second on only seven cylinders. Two time World Champion Emerson Fittipaldi made his CART debut with a 4th-place finish. Halfway through the race, Andretti's lead over second place Fittipaldi had grown to 12 laps or almost two minutes. If his helmet hadn't blocked their view, the fans in the grandstands would have probably gotten a glimpse of Mario yawning. This was probably the place for a NASCAR yellow or two to bunch the field up

"I got on the radio to Darrell (crew chief Darrell Soppe)," said Andretti, "and said, *'Talk to me, Darrell, tell me what's happening. It got pretty lonely out there."* Andretti and the No. 3

Budweiser Lola T-800 pretty much had everything their own way all afternoon, winning the 10th running of the city street race by a margin of 63.2 seconds over Geoff Brabham.

"The car was just perfect. I think today I felt I could really control the pace of anyone," Andretti told NBC Sports in victory lane. *"Once we got going at the beginning, we got an idea of who was going to be the contenders, and I felt like I could really handle it. From there on, I felt if the car stayed together we'd be alright and the car just ran absolutely perfect."*

"My car felt as good as any Formula One car ever did." The reason for that could be the fact Nigel Bennett, who worked for Lotus during Mario's successful F 1 years, designed the 1984 Lola. Unlike the "83 version, which didn't get untracked until late in the season, the T 800 appears to be a superior model. *"I was very happy Nigel came along."* continued the 1978 world champion. *"He's got a lot of experience and insight and we earned from last year's mistakes."* Brabham, whose steady, unspectacular style makes him vastly underrated in this business, turned in a fine drive in his debut with the Kraco Stereo stable. His March 84C only hit on seven cylinders all day, but he still managed to be the only other man to go the full distance. *"I thought every lap might be my last,"* said Brabham, who earned $52,300 for second. "But somehow the engine stayed together. Even though I had trouble getting past people on the straightaway, the car worked welL" In third, a big surprise considering his past record on road courses was Tom Sneva in the Texaco March 84C. *"This is the same guy who raced at Las Vegas last year"* cracked the 1983 Indy 500 winner, referring to his many off course excursions at Caesars Palace. *"This team has a lot more road racing experience and I feel that's what made the difference today."* .

Rookie Jim Crawford wound up fourth in the United Breweries Theodore, with two time world champ Emerson Fittipaldi bringing his WIT March 83C home fifth in his CART debut Most of the expected contenders were waylayed by various problems. Derek Daly, who started next to Andretti. had to settle for seventh because of fading brakes and a misfiring engine in the Provimi Veal March 84C. Teo Fabi, in the Skoal Bandit March 84C, ran second throughout the early going before spinning out. Mike Chandler was running a strong second in Dan Gurney's new Eagle Chevy before a failing gearbox ko'd him after 58 laps.

And Al Unser Jr, up as high as third but dropped out with electrical failure ii the Coors Light March 84C. There was only one actual accident when Bobby Rahal spun and collected Howdy Holmes on Lap 71. but neither driver was injured.

(All the race results can be found on the attached DVD)

While the Others Raced, Andretti Just Coasted

Fuel Supply Gamble Pays Off With a Big Victory in Long Beach Grand Prix

By RICH ROBERTS
Times Staff Writer

The Super Vee Race was won by Chip Robinson, followed by Arie Luyendyk and Steve Millen. The Toyota Celebrity Race was won by David Hobbs, also first in the professional class, followed by Scott Pruett and the first amateur, Robert Hays.

Robinson Takes Advantage of Crash With 5 Laps Left to Win Super Vee

Chip Robinson of Far Hills, N.J., took advantage of a crash involving the first two cars five laps from the finish to win the Robert Bosch Super Vee race at Long Beach Sunday.

Arie Luyendyk of Oss, Holland, was leading on the 32nd lap when Tommy Byrne of London tried to squeeze past him in the quick right-left turns at the end of the

Motor Racing / Shav Glick

Indy Cars Take Their Turns on Streets of Long Beach

When the city of Long Beach agreed 10 years ago to allow an automobile race through its streets, part of the reasoning was that a Formula One event would—along with the Queen Mary—enhance its effort to create an image as the International City.

This year the race's international aspect is gone. In place of the Formula One drivers, whose circuit takes them to many countries, will be drivers who compete in the CART/PPG World Series of Indy Car racing—as American as the Super Bowl or the Final Four.

One big question awaits Sunday's running of the Toyota Grand Prix of Long Beach, a 300-kilometer (187 miles) race on a 1.67-mile, 11-turn course around Shoreline Aquatic Park: Was it the glamour of cars like Ferrari, Renault and Lotus, sponsors like Moet & Chandon champagne and the sound of foreign languages in the pits that was the attraction—or was it just good racing?

Promoter Chris Pook is betting on the latter. In changing from Formula One to Indy cars, the Long Beach Grand Prix Assn. is saving the approximately $2 million cost of transporting Grand Prix cars, drivers and entire teams from London. Part of this saving has been passed on to the customers in reduced ticket prices.

The names of drivers are more recognizable—Al Unser, Tom Sneva, Rick Mears, Gordon Johncock, Johnny Rutherford, Danny Ongais, Kevin Cogan, et al—rather than Nelson Piquet, Michele Alboreto, Rene Arnoux, Elio de Angelis, Riccardo Patrese & Co. And English will be the language in the pits.

But, oddly, the Indy-car race does have an international aspect.

Mario Andretti, winner of the 1977 Long Beach Grand Prix and 1978 world Formula One championship as well as the 1969 Indianapolis 500 and three national Indy car championships, is in the race—a bridge as it were between the two forms of racing.

And a number of foreign drivers, including two-time world champion Emerson Fittipaldi of Brazil, have opted for the American series this year rather than the worldwide Grand Prix circuit. This group includes three who drove in last year's Long Beach race—Bruno Giacomelli of Italy, Robert Guerrero of Colombia and Danny Sullivan of Louisville, Ky.

Teo Fabi of Italy, last year's CART Rookie of the Year and winner of the final two races in 1983, is running in both circuits. Fabi failed to finish last Sunday's Formula One opener in Brazil. Other foreigners among the 41 entries are Josele Garza of Mexico, Derek Daly of

Josele Garza

Teo Fabi

And, despite this being an American race, nearly all the cars are built in England. The March cars, which account for 21 of the entries, as well as the Penskes, Lolas and Theodores are from England, and the Ligier that Cogan will drive is from France.

Practice starts at 10 a.m. Friday with qualifying sessions at 1:30 p.m. Another qualifying session Saturday at 1:30 will set the field for the fastest 24 cars in Sunday's race. Another four will probably be added at the promoter's discretion, making 28 in line for the 2 p.m. start. Also Saturday are Toyota Pro-Celebrity and the Jim Russell Pro Series races. Sunday preliminaries include the Bridgestone Pro-Kart Invitational and a Bosch Super Vee race.

MOTOCROSS—Racing promoters **Mike Goodwin** and **Mickey Thompson** are joining forces to put on a four-race stadium series in 1985 including motocross, an off-road car race, motor "extravaganza" (truck and tractor pulls, etc.) and a sprint-car and/or midget race. Goodwin is negotiating with the Coliseum, Rose Bowl and Anaheim Stadium for what he expects to be a $3- to $5-million package. The two will preview their project this year with a supercross in the Coliseum, tentatively Nov. 3. . . . Team Kawasaki's **Jeff Ward**, of Mission Viejo, season leader in both the AMA's Triple Crown Supercross series and the Bel-Ray Grand National series, gets his next test Sunday in the $25,000 Coors/Kawasaki outdoor championships at Saddleback Park in Orange. Ward also leads in the 125cc class. Other leaders are **Bob (Hurricane) Hannah** in 250cc and **Dave Bailey** in 500cc, both on Hondas. The Saddleback race also marks the return to action of a healthy **Goat Breker** of Riverside, who managed to finish third in the 500cc season opener in Gainesville, Fla., despite riding with broken ribs and a bruised

cycles Friday night at the Orange County Fairgrounds in Costa Mesa, Tuesday night at Ventura Speedway and Wednesday night at Inland Motorcycle Speedway in San Bernardino. Friday night promoter **Harry Oxley** starts his 16th consecutive year at Costa Mesa with a strong field headed by **Alan Christian**, **Mike Faria**, **Kelly Moran** and perennial champion **Mike Best**. Oxley has also landed the World Team Cup championship for 1985 at Long Beach Veterans Stadium. It will be the first time in the U.S. for the prestigious event. . . . Former world trials champion **Bernie Schreiber** of La Crescenta is tied for second place with two-time champion **Eddy Lejeune** of Belgium, two points behind **Thierry Michaud** of France after the opening round of the world championship series in Spain.

SPRINT CARS—**John Andretti**, nephew of **Mario Andretti**, will drive **Jack Gardner's** sprint car in the $30,000 Spring Nationals Friday and Saturday nights at Ascot Park. Gardner's car, driven last year by **Brad Noffsinger**, cracked the 18-second barrier on Ascot's half-mile dirt oval. Andretti, 21, was U.S. Auto Club Rookie of the Year last year and won the USAC regional championship at Indianapolis Speedrome. Also entered in the open event are three Arizonans—**Ron Shuman**, **Lealand McSpadden** and **Billy Boat**—and local favorites such as CRA champion **Bubby Jones** (who will drive earlier Saturday in a Toyota Pro-Celebrity race during the Long Beach Grand Prix), **Dean Thompson**, **Eddie Wirth** and last week's Ascot winner, **Walt Kennedy**. The win was the 45-year-old Kennedy's first win since July 29, 1979. Friday night features qualifying time trials and a 20 lap main event, with three races, climaxed by a 50-lap main event Saturday night.

OFF ROAD—It was **Stan** (not Sam) **McCuskey** who was killed in an accident during the running of the SCORE San Felipe 250 last Saturday in Baja California. Also, McCuskey was not driving (as reported in wire-service stories), but was riding with former stock-car driver **Jack McCoy** of Modesto. McCoy was taken to a San Diego hospital with chest injuries. . . . **Mickey Thompson** is the first official entry for the 17th annual Mint 400 on May 5. Drawing for starting positions in the Las Vegas-area race is Tuesday at the Mint Hotel. . . . Glen Helen Park in Devore, site of the US Festival rock concert several years ago, will host the High Desert Racing Assn.'s Nissan Classic April 7-8.

STOCK CARS—Saugus Speedway got off to its best start in the last four years with a near-capacity crowd of 5,969 for its opening race last Saturday night. **Roman Calczynski** got the win when two-time champion **Dan Press** was found to have an illegal carburetor after finishing first. Saugus is idle Saturday. The regular season opens April 7. . . . Figure 8s, oval street stocks and a chain race are on Sunday night's program at Ascot Park.

ROAD RACING—The winners of the first two IMSA Camel GT races have entered the Times/Nissan Grand Prix of Endurance April 28-29 at Riverside International Raceway. South Africans **Sarel van der Merwe**, **Tony Martin** and **Graham Duxbury**, winners of the 24 Hours of Daytona, will drive the Kreepy Krauly March 84G. Jaguars, one-two finishers in the Grand Prix of Miami, are entered, one for **Bob Tullius** and **Doc Bundy**, and the other for **Brian Redman**. . . . The Cal Club regional races, scheduled for Saturday and Sunday at Riverside, have been canceled and replaced by a bracket drag racing program.

MISCELLANEOUS—The Spring Nationals of the Hot-Rod Truck-Pull championship season is Saturday and Sunday at the L.A. Sports Arena. . . . Ventura Speedway's three-quarter midget

Michael Chandler

Jim Crawford

Bill Tempero

Ed Pimm

Roberto Guerrero

Mario Andretti

Danny Ongais

Teo Fabi

David Hobbs Takes Stage in Long Beach

British Driver Holds Off Actor Robert Hays in Pro/Celebrity Race

By RICH ROBERTS, *Times Staff Writer*

David Hobbs, wearing a large laurel wreath and waving a bottle of champagne, balanced on one leg in mock drunkenness and apologized for winning the Toyota Pro/Celebrity auto race Saturday at Long Beach.

"I'm only a lousy racing driver," he said, slurring his English accent, "not some flashy film star."

Hobbs was having fun upstaging the actor Robert Hays, who finished only 2.4 seconds behind him in the 10-lap race around the 1.67-mile course where the Indy cars will run today.

Hobbs' average speed was 56.561 m.p.h. Kart champion Scott Pruett finished second between the two, nipping Hays at the finish, and will try to win his third straight Bridgestone Pro-Kart Invitational in a preliminary event at 11 a.m. today.

Hays won the event outright three years ago but didn't seem disappointed not to hold off Hobbs and Pruett, noting the event still was fun.

He and Hobbs collected $3,000 each for winning their respective divisions, in addition to $2,000 appearance fees.

The celebrities were supposed to get a 30-second head

Toyota Celicas will require little body work for a change.

Johnson's may require a new set of tires and brakes. Unaccustomed to competing with brakes, he made the most of them by smoking his rubber into nearly every turn.

"I just go in deep and slam on the brakes," he said. "It may not show a lot of finesse, but that's what you've gotta do."

Despite his unusual style, Johnson held off sprint car champion Bubby Jones for fourth place.

Pruett's performance was important for the 24-year-old kart racer from Roseville, Calif., who hopes to move up into advanced forms of racing.

"This was the first sedan race I ever drove," he said, "and I ran right with David Hobbs, who is one of the best road racers in the world. I think I gained some recognition."

Hobbs said, "He not only drove well, he drove very cleanly. He kept his head and didn't do anything to get either of us into trouble. I managed to get Hays between us, and that definitely helped me."

Pruett is the hub of a family operation. His father

ALL ANDRETTI

The 1984 Toyota Grand Prix of Long Beach

Lola does Long Beach

Read these articles from Grand Prix International, Autoweek, Open Wheel and Racer on the attached DVD

Watch the video on our you tube channel - www.youtube.com/@racinghistoryproject2607

WELCOME TO
RECREATION · PARK
GOLF · COURSE
WE WANT YOU TO ENJOY YOUR GAME
BUT PLEASE OBSERVE THE FOLLOWING
COURTESIES FOR YOUR FELLOW GOLFERS
REPLACE YOUR DIVOTS.
GOLF COMMISSION

Canon
Canon
Canon

1985

The Toyota Grand Prix of Long Beach, held on April 14, 1984, was the opening race of the CART season. This year's schedule began on Thursday with the human powered Mini Grand Prix at Long Beach City College, the Pine Avenue Indy Car Concours d'Elegance and from 4:30 PM to 7:30 PM, the Indy Car Garage Party in the Long Beach Arena. Friday was the PPG Pace Car demonstration at 9:40 AM and 2:15 Pm and practice and qualifying all day. Shoreline Village featured a swimsuit show, entertainment and miniature Malibu Grand Prix Cars plus 50 cent draft beer.

Saturday was the Toyota Celebrity Race at 11:45 AM, more qualifying and Bosch Super Vee Race at 2:10 PM. Sunday, race day, started off with a PPG Pacecar demonstration and parade, followed by the Formula Russell Race at 10:40 AM, the Bridgestone Pro Kart Race at noon and more Toyota Pace Car and PPG Pace Car parades with the CART Race starting at 2:00 PM.

NBC TV Sportsworld broadcast the Saturday qualifying and Sunday race with Bobby Unser and Bruce Jenner as reporters and Paul Page as the announcer. KLAC radio broadcast qualifying on Saturday at 2:00 PM and the race on Sunday at Noon.

Pole sitter and last years winner Mario Andretti won in front of 67,000 spectators by calculating he could run the distance with only one pit stop for fuel and tires; a bold and risky plan by the defending champion of the race. He collected $92, 634 for the win.

A little less than two thirds into the race, it appeared that the game plan had backfired; Danny Sullivan was in the lead in the Miller American March and drawing away by almost a second a lap toward a dramatic first outing with the Roger Penske team. But Sullivan ran out of fuel twice in the last 12 laps; highly uncharacteristic for the normally flawless Penske organization; and Andretti breezed to a 60.13-second victory over Emerson Fittipaldi.

While the Others Raced, Andretti Just Coasted

Fuel Supply Gamble Pays Off With a Big Victory in Long Beach Grand Prix

By RICH ROBERTS
Times Staff Writer

When you've passed Barstow on your way to Las Vegas and your fuel gauge is flirting with empty, don't try making Baker unless Mario Andretti is driving.

The remarkable Mario could probably coast into the Strip on fumes and roll a seven before the engine died.

Sunday at Long Beach he had just enough methanol to finish the full 90 laps and claim his second consecutive Toyota Grand Prix

"We planned it along"; said Andretti's crew chief Darrell Soppe. The race was shortened to 150 miles from 187 and Andretti's crew turned the boost down. Sullivan didn't and was able to lead until running out of fuel.

Danny Sullivan's Car Simply Runs Out of Fuel

By TRACY DODDS
Times Staff Writer

A less charming guy would have been stomping around his pit pointing fingers, growling that any question about race strategy be directed to the so-called experts. That act has played the racing circuit many a season.

But that's not Danny Sullivan's style. He's more comfortable flashing that heart-melting smile, facing the not-so-comfortable facts with ease and grace.

Danny Sullivan finished the Toyota Grand Prix of Long Beach Sunday in third place because his car ran out of fuel and coasted into the pits twice.

changes last night to go faster today. It was a gamble, but it didn't work out. As you can see, a lot of others were having the same problem."

Roger Penske, owner of the March 85C that Sullivan is driving this season, was impressed that he had worked his way up from ninth place to take the lead from Andretti on one point.

"Danny didn't lose the race because of the way he drove, he lost the race because we didn't have the fuel figured out the way they did," Penske said.

"We knew we were in trouble at the first stop. That's why we turned the boost down on both cars. (The other team car was driven by

the track. His undoing was that yellow fuel light that kept winking at him while he was trying to keep his eyes on the road.

Sullivan was the first of the leaders to stop for fuel, turning in to the pits on lap 37 of the 90-lap race. "We knew we were in trouble when we had to make that first stop early," Sullivan said.

Even after making that early stop, Sullivan worked his way back into second place before Andretti made his one and only stop—but Andretti had such a big lead that he was still in the lead when he came back out.

In order to run the entire race with just one stop, Andretti dialed his boost back to conserve fuel

Andretti pass.

In fact, he let Andretti pass only after Sullivan had passed Andretti on turn 9 of lap 59.

Unser, of course, denies intentionally blocking Andretti. Sullivan said he heard nothing about such a fiendish plan on his radio. "I don't think Al chopped him any more than he would have whether I had been behind him or not," Sullivan said.

Sullivan enjoyed the lead for 20 laps, stretching it to more than 16 seconds before coasting into his pit with an empty fuel tank.

"That was what killed us," Sullivan said. "We had to come in with a head motor. Then we had a problem with the starter motor and may

Al Unser Jr/ won the Toyota Celebrity Race and the professional class, followed by Parnelli Jones and Dan Gurney. First in the amateur class was Lorenzo Lamas back in eighth. The media class was won by Tony Swan who was fourth. The highlight of the race was Jackie Jackson's crash and injury. Davy Jones won the Bosch Super Vee Race followed by Ken Johnson and Ted Prappas. Lynn Haddock won the Bridgestone Kart Race and Lat Franklin won the Formula Russell Race.

(See all the race results on the attached DVD)

It's Somewhat of a Thriller for Jackie Jackson

Oldest Member of Singing Group Wrecks Car, Hurts Leg in Pro/Celebrity Race

By RICH ROBERTS,
Times Staff Writer

The next time somebody tells singer Jackie Jackson to "break a leg," maybe he'll take them less seriously. Or stay out of race cars.

The oldest member of the Jacksons, the slowest qualifier for the Toyota Pro/Celebrity race at Long Beach Saturday, wrecked his second car in two days. He was taken to St. Mary Medical Center with a possible fractured right leg, but X-rays were negative and he was released.

Jackson was accompanied by his brother Randy. The best-known Jackson, Michael, was not present.

Jackson's publicist, Debbie Baum, said the injured singer was "laughing and giggling when they pulled him off the stretcher. He had so much fun you'll have trouble keeping the other brothers out (of the event) next year."

It was the same leg on which Jackson had knee surgery before

JOE KENNEDY / Los Angeles Times

Jackie Jackson is helped from his car after Long Beach crash.

"He braked late or didn't downshift," she said. "He slammed into it

brother. He knows these (Super Vee) cars inside and out."

easily lose track of where you are. It's particularly hard to tell your braking points. You have to be careful not to daydream."

□

It Pays To Advertise: Most of the race cars carry more advertising than the late-night movies, but some still have to hustle sponsors.

On the side of the Super Vee owned by Giupponi Franca of Hollywood is a sign: "THIS SPACE FOR RENT."

□

Dept. of What Are You Doing Here?: Cleveland Browns' linebacker Tom Cousineau follows the Indy-car circuit as a driver for the PPG pace car team.

Last year he worked on Scott Brayton's pit crew. His position: "Left rear."

□

Team Esprit, a two-women-power entry in Little Long Beach Grand Prix, gets away first from starting line in preliminary heat.

Bruno Giacomelli leads Geoff Brabham

Sports Illustrated

Good to the very last drop

Watch the video on our you tube channel - www.youtube.com/@racinghistoryproject2607 and read the Sports Illustrated and Indy Car Magazine article on the attached DVD

Michael Andretti

Geoff Brabham

Danny Sullivan

Mario Andretti

Garza, Fittipaldi and Ongais

Al Unser Jr. and Jacques Villeneuve

Dan Gurney and Lorenzo Lamas

1986

The second race on the CART calendar, the Toyota Grand Prix of Long Beach, was held on April 13, 1986 with a race day crowd of 77,500. The race was 95 laps, trying to avoid the economy run of previous years.

A busy week scheduled; the PPG Grand Prix Celebrity Golf Tournament the week before then on Thursday at 11:00 AM, the Mini Grand Prix at Long Beach City College and at 2:00 PM,, the Marlboro Pit Stop Competition downtown followed by the Committee of 300 Garage Party at 4:30 PM.

Friday was practice and qualifying. On Saturday, there was the Toyota Celebrity Race at 11:45 PM followed by the Super Vee Race at 2:10 PM. On Sunday, the Jim Russell Pro Series ran at 10:40 AM, and new this year, an IMSA GTP exhibition race at 12:30 PM and the CART Race at 2:00 PM

Danny Sullivan had the pole, followed by Al Unser Jr, and Mario Andretti but the race became a battle of second generation drivers as 23 year old Michael Andretti kept the Andretti name synonymous with Long Beach by winning a bumper to bumper, tooth and nail confrontation with Al Unser Jr. over the final 24 laps of the race.. Geoff Brabham was third. The lead changed hands six times among five drivers but, with both Andretti and Unser making early fuel stops, Andretti took the lead for the final time after Unser pitted for fuel on lap 69 and won by .38 seconds. Andretti collected $94,546 for the win.

Danny Sullivan quickest both on and off Grand Prix course

By Jim Thomas
Staff writer

Danny Sullivan, the colorful winner of last year's Indy 500, was at his best both on and off the track Friday during first-day qualifying at the Toyota of Long Beach Grand Prix.

Not only did he turn in the day's fastest time, he also turned in the best off-the-cuff remarks.

Asked why he wasn't wearing any socks at the interview stand, Sullivan said "they don't pay me enough. I can't afford them.

"No, really, I have sponsored socks, but they're being washed. You think I'm kidding, but I'm not.

"We have so many drivers on our team, we all have to have monogrammed socks so the guys who hand out the equipment will know what to do."

Asked about a slight change in turns seven and eight that slowed down the qualifying runs, Sullivan

car after his March 86C blew an engine during a practice run.

Like many other drivers, Cogan hopes to do better in today's final qualifying run for Sunday's race.

"All things considered, eighth is pretty good," Cogan said. "With more time to get it together, we could move up tomorrow."

Sullivan is well aware of that.

"No, I won't hold anything back," he said. "If you wait to see how everyone else is doing, you just might get caught napping.

"Times will be faster tomorrow if the weather remains good. It was almost perfect today — cool and overcast, but no winds.

"If it rains, though, the times might hold up."

If it does, it would be the fourth time Sullivan had earned the pole position in his CART/Indy Car career.

"Sure, I want the pole position real bad," Sullivan said with a smile. "Like I said, it's going to

Look Out, Here Comes the New Generation

Sons of Famous Drivers Finish 1-2-3 in the Grand Prix of Long Beac[h]

By RICH ROBERTS
Times Staff Writer

Al Unser Jr. didn't even wait to be asked the obvious question.

"I want to say that the changing of the guard has happened," he said after finishing less than a half-second behind winner Michael Andretti in the Toyota Grand Prix of Long Beach Sunday.

With Geoff Brabham, 34, placing third, the theme could have been from "My Three Sons": all offspring of world and/or Indianapolis driving champions, a new generation trying to claim racing success in its own right.

Unser, especially, about to turn 24 next Saturday, is brash, young and fast. He has won three CART Indy-car races in five years and

Jeff Krosnoff Edges Jon Beekhuis to Win Mazda Series 20-Lap Race

Jeff Krosnoff, 21, passed his friend Tom Kendall, 19, on the fourth lap and went on to win the 20-lap Russell/Mazda Pro Series race by 2.16 seconds over Jon Beekhuis at Long Beach Sunday.

It was Krosnoff's first victory in three years in the series for open-wheel cars powered by Mazda rotary engines. He and Kendall are La Canada neighbors who attend UCLA.

Later in the race, a prelude to the Toyota Grand Prix of Long Beach, Kendall spun on Turn 5 and lost a place to Beekhuis, 25, of Salinas, and finished third.

Beekhuis, the chief instructor at the Jim Russell British School of Motor Racing at Laguna Seca, said he learned something from Krosnoff and Kendall.

"They had a trick in the hairpin to carry more speed through the corner," he said. "It took me a while to pick it up."

—RICH ROBERTS

could go on for another quarte[r] century.

"I don't know, Michael's going [to] be something to contend with f[or] years to come, and I'm going to [be] in it for a long time," Unser sai[d]. "I love racing."

But Andretti wasn't so sure th[ey] have heard the last of the eld[er] generation. His first CART win [in] four years on the circuit kept t[he] Long Beach event in the family f[or] the third consecutive year, but [he] didn't think it represented a trans[i]tion.

"I still feel the guard has[n't] changed," he said. "It's gonna [be] tough to keep doing this. I thi[nk] things just went our way. Ther[e're] gonna be days when things go da[ng] way, and he'll be back up the[re] again."

Early in Sunday's race, t[he]

Dan Gurney won the Toyota Celebrity Race. Second was Parnelli Jones and third, first of the amateurs, was Perry King. Steve Bren won the Super Vee Race, followed by Mike Groff and Jeff Andretti. Jeff Krosnoff won the Pro Russell Race, followed by Jon Beekhuis with Tom Kendall in third.

JAYNE KAMIN / Los Angeles Time[s]

Michael Andretti (center), Mario's 23-year-old son, celebrates wi[n] with third-place Geoff Brabham (left) and runner-up Al Unser Jr

(See all the race results on the attached DVD)

Associated Press

Al Unser Jr. leads pack around the first turn at Long Beach Grand Prix. Michael Andretti overtook Unser on 69th lap en route to victory.

Photo by PETER JAMES SAMERIJAN

Raiderette cheerleader Cindy Sullivan was crowned 1986 Miss Toyota Grand Prix of Long Beach during a sold-out ceremony Feb. 28 at the Golden Sails Hotel in Long Beach. Cruise and airline tickets were among the prizes won by the 23-year-old Huntington Beach resident, who was selected from among a field of 20 contestants.

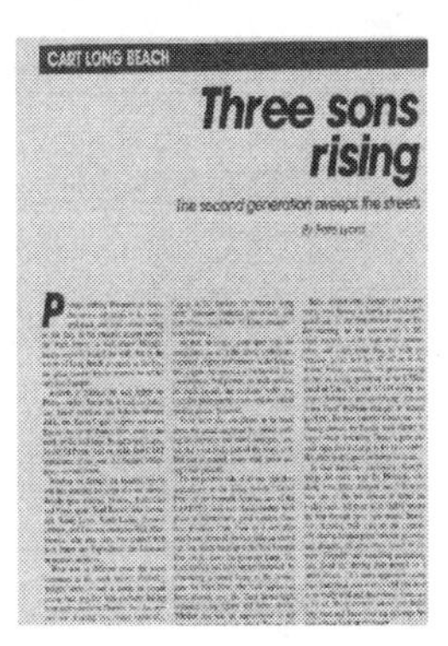

Watch the video on our you tube channel - www.youtube.com/@racinghistoryproject2607 and read the Indy Car and Autoweek articles on the attached DVD

Al Unser Jr,

Emerson Fittipaldi

Tom Sneva

Robert Guerrero

Johnny Rutherford

Raul Boesel

Josele Garza

Jan Lammers

Marlboro
Marlboro
Marlboro
Marlboro
Marlboro
Marlboro
SCOTSMAN

TOYOTA
Budweiser
7-ELEVEN

1987

The 1987 Toyota Grand Prix of Long Beach, the first race of the CART season, was held on April 5, 1987. 82,500 people saw Mario Andretti win for the third time in Indy cars (he had a win in Formula One also). Other races included the Bosch Super Vees, Bendix Trans Am, new at Long Beach this year and the Toyota Celebrity Race.

Many events during race week; with the Wine and Dine d'Elegance the Saturday prior; food sampling, wine tasting and classic cars at Shoreline Aquatic Park and Sunday's 8:00 AM 10K run at the Queen Mary and the all day Tecate Grand Prix Western Festival with a Chili Cook off, Ugly Dog Contest and Valvoline Mini Grand Prix.

Tuesday was the PPG Grand Prix Charity Golf tournament at Recreation Park and on Wednesday, the Black Tie Charity Ball at the Naval Officers Club. Thursday was the Mini Grand Prix at Long Beach City College at 11:00 AM, the Downtown Long Beach Racing Car Concours at 12:30 PM and the Long Beach Motorsports Expo and Committee of 300 Garage Party, beginning at 4:00 PM.

On Friday, the Motorsports Expo was open all day while there was practice and qualifying on the track. Saturday had more practice and qualifying for all groups plus the Toyota Celebrity Race at 1:00 PM and the Bendix Trans Am Race at 4:00 PM. Sunday, race day, started with the PPG Pace Car on track exhibition at 9:45 AM followed by the Super Vee Race at 10:40 and the CART Race at 2:00 PM.

Not Only Does Mario Win Pole, He Owns It

By SHAV GLICK, *Times Staff Writer*

Since the first Long Beach Grand Prix in 1975, when cars raced along Ocean Boulevard and then dipped down steep Linden Avenue before sweeping along Shoreline Drive in front of the Queen Mary, there have been many changes.

it. He is going like a rocket."

It took Andretti 65.886 seconds to negotiate the Long Beach circuit in his tomato-red Hanna Auto Wash Lola. Guerrero took 66.806. That margin, if extended over the 95

Mario Andretti won in the Paul Newman / Carl Haas Ilmor powered Lola and lapped the entire field, followed by Al Unser Jr, and Tom Sneva. Mario, starting from the pole. led all the way and collected $97,410 for the win.. Behind him, son Michael had a incident with Roberto Guerrero at the start and Ist time replacing the front wings. Tom Sneva had wastegate issue but was able to hang on for third. Fourth place Josele graza, still recovering from last years crash at Mid Ohio, was happy to record a finish in only his second race since the accident. Bobby Rahal tagged the wall on his sixth lap. Danny Sullivan was running in the top ten when he had an ignition failure Emerson Fittipaldi ran a close second to Mario until a waste gate failure put him out. Only 13 cars finished as the rough city streets took their toll.

Mario Andretti wins easily

Family continues domination of Long Beach Grand Prix

The Associated Press

LONG BEACH — With his car running "picture perfect," Mario Andretti drew away from the rest of the field yesterday in the Toyota Grand Prix of Long Beach, the Indy-car season opener.

Andretti, continuing the domination by his family of this picturesque seaside downtown-streets event, had only former

MOTOR SPORTS

charger failure, Andretti found himself in complete control.

Behind Unser was Tom Sneva, followed by the younger Andretti and Josele Garza of Mexico, both two laps off the pace.

Andretti's average speed of 85.330 mph broke his son's year-old Long Beach record of 80.968.

grabbed an early lead and went on to his second straight victory in the SCCA Bosch-Volkswagen Super Vee race, also through the streets of Long Beach.

Pole-winner Paul Radisich of New Zealand finished second.

Bren averaged 79.911 mph and picked up first-place money of $6,500.

Earnhardt NASCAR victor

Steve Bren won the Super Vee race with Paul Radisich second and Scott Atchison third. Jason Bateman won the Toyota Celebrity Race and the amateur class with Richard Dean Anderson in second. Juan Fangio, the first professional, was third. Scott Pruett won the Bendix Trans Am Race with Tom Kendall second and Bruce Jenner third.

Charity Events, Race Party Help Hype the Grand Prix

By ROXANA KOPETMAN, *Times Staff Writer*

LONG BEACH—The temporary bleachers and concrete barriers are in place as downtown Long Beach, once again, is transformed into a serpentine 1.67-mile race course for this weekend's Grand Prix activities.

"The city becomes a race circuit going on: It's Grand Prix time."

"People start driving Ocean Boulevard a little faster because it's part of the circuit," Green said.

The circuit runs from Shoreline Drive to Seaside Way, cuts under the Hyatt Regency garage, contin-

Saturday's Races

King Hits Wall, and Win Goes to Bateman

By RICH ROBERTS, *Times Staff Writer*

LORI SHEPLER / Los Angeles Times

Disc jockey Hollywood Hamilton helps send Perry King, the defending champion, into the wall and out of the pro-celebrity race on the Long Beach Grand Prix course Saturday.

Perry King, star of adventure movies and television series, was saying just the other day how Hollywood has been carried away with scenes of cars crashing and burning.

"I'm not really into that," King said. "I think it's overdone."

So guess who crashed and burned in the Toyota Pro-Celebrity Race on the Grand Prix circuit at Long Beach Saturday.

King, who won the event last year, shrugged and said, "I don't think it really burned. It was just smoking. Somebody punted me into the wall."

King's culprit appeared to be Chad McQueen, who checkmated King into the wall on the Seaside Way back straightaway. The son of the late Steve McQueen, who appears in "Bullitt II," also will drive in today's Bosch Super Vee race at 10:40 a.m.

The 17 entries, including seven professionals who started the race in a pack 30 seconds behind, played 10 laps of Destruction Derby before 18-year-old Jason Bateman of "Valerie" emerged the winner by a fraction of a second—.646—over Richard Dean Anderson of "MacGyver."

Pro Juan Manuel Fangio II was another 2.256 seconds behind.

King's car, its front end destroyed, had to be brought in by a wrecker. Fangio tapped retired Angel Bobby Grich into a spin while passing him on the hairpin turn, and disc jockey Hollywood Hamilton finished on three wheels with debris from his car strewn around the track.

Scratch a couple of Celicas.

"Hamilton hit me twice," said "Leave It To Beaver's" Tony Dow.

KABC-TV's Ted Dawson: "Hollywood spun me out in Turn 1."

"Dynasty's" Ted McGinley: "Holy [bleep]! Who let Hollywood Hamilton on the course? Incredible!"

Hamilton left the circuit after the race and was unavailable to explain his driving style.

The venerable Dan Gurney admitted he "made a mistake. My brakes locked and I went off the escape road at Turn 9."

Bateman is the youngest winner in the event's 11 years and the first celebrity to finish first since Olympic decathlon champion Bruce Jenner in 1982. It was the second closest finish next to Robert Hays' win over Parnelli Jones by 0.34 in '81.

Anderson, the polesitter, led the race for five laps, with Bateman in his draft. Then, moving between Turns 1 and 2 at the end of Shoreline Drive, Anderson's car jumped out of first gear and when he slowed Bateman tapped him from behind.

"He hit my rear end and went on past and that was it," Anderson said. "Then I was on his tail the rest of the way."

Bateman said: "I hope it wasn't because of that little tap. I bumped him and he veered off to the right. I just floored it in second and went past him."

Fangio, 30-year-old nephew of Argentina's Formula One legend of the 50s, is campaigning in the ARS in the U.S. but took Saturday's race seriously.

"He put me on an E-ticket ride," Grich said.

Fangio: "He was going a little slow. In all types of racing, everybody wants to win."

Or at least survive.

□

Scott Pruett of Roseville, Calif., showed that he knows how to win from either end of the field when he won the season-opening Bendix Trans-Am race by 1 minute 2.33 seconds over Tom Kendall, 20, a UCLA junior.

Pruett, 26, led from the pole for 19 of the 60 laps before moving over to let Roush racing teammate Peter Halsmer by. Then when Halsmer dropped out with a broken connecting rod after 39 laps, Pruett cruised to victory.

A year ago Pruett missed qualifying at Riverside and started his first Trans-Am race from 31st position in the last row, but picked his way through the pack to win.

"This victory is real sweet," he said, "but it's a little sweeter coming from the back."

A third Roush entry, Deborah Gregg of Jacksonville, Fla., started 28th in her Trans-Am debut but placed eighth, two laps behind Pruett in the strung-out field.

Five laps (40-45) were run under a yellow caution flag when a plugged rain gutter at the Hyatt Hotel suddenly opened and flooded Turn 7 near the hotel garage.

(See all the race results on the attached DVD)

Shoreline INN
ANAHEIM
mazda
SHANNON Corp
simple green
Donnelley
Marlboro
WELCOME RACE FANS
GOODYEAR
DOMINO'S PIZZA
DOMINO'S PIZZA
Coke
VALVOLINE
VALVOLINE
BOSCH Spark Plugs
BOSCH Spark Plugs
MACHINISTS
MACHINISTS
55
Schaefer
Schaefer

Tom Sneva

Bobby Rahal

Roberto Guerrero

Emerson Fittipaldi

Mario Andretti

Michael Andretti

CART LONG BEACH

Heartbeats of America

Grandpa Andretti gives the Chevy V8 its first Indycar win

Watch the video on our youtube channel - www.youtube.com/@racinghistoryproject2607 and read the Autoweek article on the attached DVD

Bruce Jenner – Third in the Trans Am Race

Andretti, Unser and Sneva

1988

The 1988 Toyota Grand Prix of Long Beach, round two of the CART championship, was held on April 17, 1988 in front of a crowd of 84,000. On the Saturday before, a Wine and Food Tasting and Concours d' Elegance were held in Shoreline Park. Sunday morning featured a 10K run at the Queen Mary and the all day West Fest with a Chili Cook Off, a Mini Indy Race, Miss Chili Pepper and Mr. Hot Sauce contests, a Pig Race and the Ugly Dog Contest..

Tuesday was the Charity Golf Tournament at 9:00 AM, Wednesday the Grand Prix Charity Ball at 6:30 PM and Thursday the Concours d'Promenade at 11:30 AM followed by the Committee of 300 Indy Garage Party at 4:00 PM.

Racing events began on Friday with the opening of the Motorsports Expo and practice and qualifying all day. Saturday was more practice and qualifying with the Toyota Celebrity Race at 1:00 PM with the Trans Am Race at 4:00 PM.

The Super Vees raced at 10:15 AM on Sunday followed by the CART Race at 1:00 PM. The race was broadcast on ABC, blacked out in the Los Angeles area. with reporters Bobby Unser, Sam Posey and Jack Arute. Paul Page was the lead announcer.

Qualifying

Sullivan Gets Pole Position

By SHAV GLICK,
Times Staff Writer

LORI SHEPLER / Los Angeles Times

Course workers signal to other drivers after Scott Brayton lost control of his car and hit the wall during Saturday's qualifying round in Long Beach. Brayton was not injured in the mishap.

Danny Sullivan found a clear stretch of race track for 1 minute 6.607 seconds Saturday afternoon and in that short span of time wrested the pole position from Mario Andretti for today's Toyota Grand Prix of Long Beach.

Andretti, winner of the Long Beach pole in three of the past four Indy car races, had one last chance to get No. 4 but as he came up on the final hairpin turn leading to the start-finish line, he found Scott Pruett in the turn. In the blink of an eye that it took Andretti to get around Pruett, he had lost a tenth of a second and the pole.

Sullivan turned a lap around the 11-turn, 1.67-mile seaside circuit at 90.261 m.p.h. to win his second pole for the Long Beach race. He was also on the pole in 1986 with a speed of 90.318 m.p.h.

Andretti's best was 90.113, good enough to put him on the front row alongside Sullivan when the race is flagged off at 1 p.m.

"It sounds like a bad excuse, but I got caught up in traffic twice," Andretti said.

Sullivan was driving a Chevrolet-powered Penske PC-17, the stablemate of the car Rick Mears drove to win the pole last week in the Checker 200 in Phoenix—a race that Andretti won in a Chevrolet-powered Lola.

It was Sullivan's sixth career pole in Indy cars and his first since Cleveland in 1986, which was also his last win.

All but five of the 26 cars bettered their times from Friday on a cool, cloudy day ideal for qualifying. Twenty-six will start the 95-lap, 158.65-mile race, 24 from qualifications and Dominic Dobson and Dick Ferguson by promoter's option.

Al Unser Jr. won, led all but 22 laps and pocketed $91,160. He had a lap and a half lead over second place Bobby Rahal with Kevin Cogan in third. After following Mario Andretti at the start, he passed him for the lead at the end of lap one. He fell back to sixth after a pit stop problem but recovered. Pole qualifier Danny Sullivan finished 13th. Last year's winner Mario Andretti had clutch failure and son Michael was seventh. Rookie Scott Pruett ran as high as sixth before his engine gave up. Scott Brayton brought out the only yellow flag of the race when he spun in the hairpin.

In the supporting races, Paul Gentilozzi won the Saturday Trans Am with Hurley Heywood in second and Lyn St. James in third. The race was red flagged for 43 minutes due to a bad crash and serious injury to Dan Croft who later died. Paul Moyer of KABC TV won the Toyota Celebrity Race and the amateur class, beating professional Dan Gurney. Former 49'ers player Dwight Clark was third. The Super Vee Race was won by Ken Murillo with Paul Radisich in second.

Little Al Writes a Flying Finish to Andretti Rule

By SHAV GLICK
Times Staff Writer

All week long, Al Unser Jr. said all he wanted was to move up one position in the race through the streets of Long Beach.

Little Al had finished second two years in a row, to Michael Andretti in 1986 and Mario Andretti in 1987.

Sunday he got what he wanted.

Unser, who will be 26 on Tuesday, thoroughly dominated his opposition in winning the 14th annual

A crowd estimated at 84,000, plus thousands more who were watching from balconies of high-rise buildings, or eavesdropping from a fleet of ships from the Queen Mary to the the marina, witnessed the colorful spectacle on a cloudy, chilly day.

Only a lengthy pit stop, caused when a wheel wouldn't come off Unser's March-Chevrolet during a tire change, prevented the race from being a boring runaway.

Unser, who won by one lap and 33.48 seconds, knew it was going to

(All the race results are on the attached DVD)

DAVID EULITT/The Sun

TOPPING OFF: Al Unser Jr. sits inside his March-Chevrolet vehicle as helmeted crew members pump gas during a pit stop on Lap 86. Unser had a comfortable lead at the time, and went on to win the Long Beach Grand Prix for Indy cars.

LORI SHEPLER / Los Angeles Times

Marlboro
Marlboro
20
Mobil
BOSCH

STROH
3
STROH
3
Valvoline
Valvoline

Trans Am Winners - Gentillozzi, Heywood and St. James

Trans Am Action

Al Unser Jr., left, who won the Toyota Grand Prix of San Diego Sunday, gets a champagne bath from Kevin Cogan, right, who finished third. (AP Laserphoto)

Emerson Fittipaldi

Al Unser Jr.

AP Laserphoto

Indy-car driver Danny Sullivan set the fastest qualifying lap for today's Long Beach Grand Prix at 90.261 mph.

Some racers in Pro-Celebrity event, from left, Walter Payton, Susan Ruttan, Jason Bateman, Richard Dean Anderson.

Celebrities in Grand Prix Driver's Seat

By PAUL DEAN, *Times Staff Writer*

Round up the usual number of stars, goes the public perception, get them at least to look their part in helmets, fireproof driving suits and cars with roll bars—and then let them pretend to be fast and tough for 10 slow and cushy laps of today's Pro-Celebrity preliminary to the Toyota Grand Prix of Long Beach.

"But that's not quite it," corrected Les Unger, motor-sports manager for Toyota and guardian of this 1 p.m. warm-up to the Grand Prix proper and Sunday's entry of such gladiators as Mario Andretti, A. J. Foyt, Bobby Rahal and Emerson Fittipaldi. "This event is not a giggle for any of the celebrities.

"It is serious competition for those who want to take it to that level. . . and I would say that half of the celebrities in today's event will be running at 110%."

In fact, one of Unger's favorite stories when challenged about race hype versus actuality concerns the intimidation of an athletic giant after tackling 100-m.p.h. speeds between the concrete barricades of the 1.6-mile Long Beach circuit.

"This terror of the National Football League, Mark Gastineau, went out to practice against Tony Danza, came back in and said to me: 'This is scary stuff. This is worse than playing defensive end and getting blocked out by six or eight guys.' "

Celebrities have been injured during the event. The most publicized pileup was that of actor Lorenzo Lamas, who made the cover of National Enquirer (and in color) after totaling a Toyota and breaking his foot during the 1985 event.

And that was while *training* for the race.

Actor Broke Arm

Singer Jackie Jackson hit a wall that same year and was hospitalized for observation. Woody Harrelson broke a bone in his arm when his car flipped in practice for last year's event—which explains one of the casts (the other was the result of a fist fight) he wore in last season's episodes of "Cheers."

Accidents, Unger acknowledged, do happen, "but we continue to pride ourselves on the amount of training/preparation of cars and celebrities."

Once invited to compete, he said, all contestants must spend four days at the Drivers Connection, a high performance driving school at Willow Springs Raceway, near Rosamond.

"We're not trying to make race-car drivers out of these people," Unger said. "We're teaching them the basics of car control, how to shift, how to keep engine revolutions at the right level, how to enter and exit a turn and how not to squeal your tires, because all that stuff that sounds neat doesn't make you go any faster.

"At the end of that period, their instructors will let us know what they think of an individual's physical skills and mental outlook. If they say no to a person, as they have in the past, then those people do not race."

The celebrities drive identical cars—the 16-valve, Toyota Celica GT-S—that are unmuffled stock, plus safety devices.

"We spend a lot of time preparing these cars to make them as safe as possible," Unger said. "That's why we have roll bars, fire extinguishers, safety netting over the driver's window . . . and those instructors carefully watching everyone right up to the time of the race."

There will be 17 racers in this year's field—three professional drivers and 14 celebrities, all competing for the $5,000 first prize in both divisions. Second placings are worth $4,000, third $3,500 and fourth $3,000, with also-rans receiving $2,000 for simply showing up and driving.

To keep everything in balance, there will be a handicap start with the professional drivers 30 seconds behind the celebrities.

Please see RACE, Page 4

COMPETITION

CART Toyota Grand Prix, Long Beach, Calif.

Little Al Makes His Own Brakes

Dramatic move in hairpin propels him to dominating Indycar victory

By Gordon Kirby

Watch the video on our you tube channel - www.youtube.com/@racinghistoryproject2607 and read the Autoweek article on the attached DVD

1989

The 1989 Toyota Grand Prix of Long Beach, round two of the CART Championship, took place on April 16, 1989 in front of a crowd of 85,000. The week started off with the PPG Charity Ball on Wednesday. Thursday had the Long Beach City College Grand Prix at 11:00 AM, the Concours d'Promenade at 11:30 and the Committee of 300 Indy Car Garage / Motorsports Expo Preview at 4:00 PM.

Friday was filled with practice and qualifying. Saturday, along with Indy car practice and qualifying, the Toyota Celebrity Race was scheduled for 11:00 AM with the Trans Am Race at 4:15 PM.

Sunday, race day started off with the conclusion of the One Lap of America at 9:30 AM, the HFC sponsored American Race Series event, a feeder series to Indy car, at 10:00 AM, the CART Race at noon and the Toyota Formula Atlantic Race at 2:35 PM. ABC broadcast the race live but it was blacked out in the Los Angeles area. A tape delayed telecast will be shown at 3:00 PM. Commentary was provided by Bobby Unser, Sam Posey and Jack Arute while paul Page was the announcer.. ESPN broadcast the Toyota Celebrity Race on April 19th and the Trans Am Race on April 25th.

Unser holds pole in Long Beach GP

TV: 4 p.m., ABC Ch. 9

FLORIDA TODAY Wires

LONG BEACH, Calif. — Al Unser Jr. came up with the first pole of his seven-year Indy-car career Saturday, holding the top qualifying spot for the Long Beach Grand Prix despite failing to improve on his first-day qualifying speed.

Al Unser Jr. had the pole, next to Michael Andretti, jumped out to an early lead and led 72 of the first 74 laps but late in the race, Unser found himself in a battle with Mario and Michael Andretti. All three drivers made their final pit stops, and after a faster pit stop, Mario Andretti emerged as the leader on lap 78, with Unser second, and Michael now a distant third. Unser was close behind Mario when they approached the lapped car of Tom Sneva. At the exit of turn two, and going into turn three, Unser dove under Mario Andretti for the lead, but punted Mario's right rear wheel. Andretti was sent spinning out with a flat tire and broken suspension, while Unser broke part of his front wing, and bent his steering. Despite the damage, Unser nursed the crippled car to the finish line, winning by 12.377 seconds over Michael Andretti and collecting $117,660 for the win.

The contact was controversial, and after the race, Mario called the move "stupid driving." Unser accepted blame for the contact. Andretti, livid with anger, stormed into the impound area where Unser Jr.'s car was parked for post race inspection and said in sarcastic overtones: *"Thanks a lot. I really appreciate it. I hope we can do it again sometime."* As the top three finishers were driven around the track in a Toyota pick-up truck for a victory lap, Unser was greeted by boos, catcalls and rude gestures. *"I tell you what, that was an unpopular win,"* Al laughed. *"We went around in the pace truck and we got over to that turn two, turn three area and man, there was a bunch of boos. They weren't happy about me dethroning Mario the way it happened, and neither was I."*

Associated Press

Eventual winner Al Unser Jr. (2) leads Mario Andretti (6), with whom he later collided, into the first turn of Sunday's Long Beach Grand Prix.

Unser Jr. wins, but Andretti isn't pleased

(See all the race results on the attached DVD)

Ricky Schroeder won the Toyota Celebrity Race and the amateur class with professional Parnelli Jones in second and Dan Gurney in third. Pole sitter Tommy Byrne won the HFC American Race Series event followed by P.J. Jones and Mitch Thieman. Irv Hoerr won the Trans Am with Tom Kendall in second and Robert Lappalainen in third.

Lap of America 10,000-mile road rally will conclude at race

The One Lap of America, a non-stop road rally around the United States, got the green flag Thursday on the main straightaway of the Long Beach Grand Prix course.

A field of at least 60 cars—from vintage machines of the 1930s to the latest imported and domestic models—began the 10,000-mile, non-stop trek in last week's wilting heat wave.

The finishers are due back on Saturday. The top five finalists will compete on the Grand Prix course 9:30 a.m. Sunday to determine the winner.

The sixth lap of the U.S. map is the first to cover a 10,000-mile route in 10 days. The first five were held on an 8,000-mile course.

The National Multiple Sclerosis Society has been named the official beneficiary of the 1989 One Lap of America.

Toyota Motor Sales U.S.A. Inc. will provide specially equipped Celica All-Trac turbocharged hatchbacks for event officials on the grueling journey, which will take competitors over the Rocky Mountains twice and to such challenging tracks as Indianapolis Raceway Park, Sebring International Raceway in Florida and the famed climb to the sky at Pike's Peak, Colo.

COMPETITION

It's the Family Feud

Watch the video on our you tube channel - www.youtube.com/@racinghistoryproject2607 and read the Autoweek article on the attached DVD

ICHELOB
BELL
parmalat
Marlboro
5
16

Marlboro
TOYOTA
STROH'S
Marlboro
STROH'S
TOYOTA

Mario Andretti

Rick Mears

Scott Pruett

Danny Sullivan

James Weaver

Teo Fabi

Raul Boesel

Al Unser Jr.

Kevin Cogan

Michael Andretti

John Jones

Emerson Fittipaldi

TOYOTA
#1 DEALER IN THE U.S.A. FOR 22 YEARS
COME SEE WHY!
LONGO
LONGO GRAND PRIX
WE'RE OUT TO SELL 2000 TOYOTAS IN APRIL!
$0 DOWN
ON APPROVED CREDIT
ABSOLUTELY NO CASH
REQUIRED TO PURCHASE ANY OF OUR NEW TOYOTA CARS OR TRUCKS IN STOCK
FINANCE REPRESENTATIVES ON THE PREMISES FOR IMMEDIATE CREDIT APPROVAL!
LONGO'S WINNING FORMULA
•SERVICE
•SELECTION
•PRICE
LONGO TOYOTA IS RACING FOR A RECORD,AND TO WIN WE MUST GET YOU AN UNBEATABLE DEAL ON A BRAND NEW TOYOTA CAR OR TRUCK.
TOYOTA SET THE PACE, AND LONGO CAN HELP YOU WIN THE RACE, WITH
2000
NEW TOYOTAS IN STOCK FOR IMMEDIATE DELIVERY!
RICK MEARS
DANNY SULLIVAN
AL UNSER
PENSKE RACING
GOOD LUCK to THE MOST SUCCESSFUL TEAM IN INDY CAR HISTORY
1989 TOYOTA GRAND PRIX OF LONG BEACH
THE TEAM PENSKE FRONT ROW AT THE 1988 INDY 500
PENSKE RACING
AMERICA'S #1 DEALER
GIANT
LONGO
A PENSKE COMPANY
TOYOTA
AT THE 125 FT LONGO SIGN
10 FWY PECK RD NORTH EXIT
VALLEY BLVD.
PECK ROAD
LONGO TOYOTA
3534 NORTH PECK ROAD., EL MONTE (213) 686-1000 (818) 580-6000 (714) 556-7563

1990

The Toyota Grand Prix of Long Beach was held on April 22, 1990 in front of a crowd of 80,000. Monday night was the Grand Prix Kickoff Party, 5:00 PM at the World Trade Center. Thursday was the Concours d'Promenade, downtown between Ocean and Third Street and Friday was practice and qualifying for everyone as was Saturday with the Toyota Celebrity Race scheduled for 1:00 PM and the Chevron GTO / GTU Challenge Race at 4:15 PM.

Sunday featured the American Racing Series event at 9:45 AM, the finish of the One Lap of America at 11:00 AM, the main event, the CART Race, at 1:00 PM and the Toyota Atlantic Race at 3:30 PM. The race was broadcast on ABC TV, tape delayed in Southern California. Commentators were Bobby Unser, Jack Arute and Sam Posey. Paul Page was the announcer.

Unser Jr. Qualifies According to Past Form

By SHAV GLICK
TIMES STAFF WRITER

There seems to be something special about the Long Beach street course as far as Al Unser Jr. is concerned.

Little Al, whose father Big Al has won four Indianapolis 500s, is on a streak. He has won the last two Toyota Grand Prix of Long Beach races in the PPG/CART Indy car series. Last year, he also won his first Indy car pole on the 11-turn, 1.67-mile temporary circuit.

Friday, in the first day of qualifying for Sunday's $1-million race, he was again fastest.

Unser, 28 Thursday, drove his Lola T9000, powered by a Chevrolet Indy V-8 engine, at an average speed of 89.196 m.p.h. He had qualified at 90.740 last year.

MINDY SCHAUER / Los Angeles Times

Penske PC-18 that Emerson Fittipaldi drove last year when he won the Indianapolis 500 and the PPG Cup driving championship.

Asked to contrast Indy cars and Grand Prix cars, Cheever said: "When you're driving, it doesn't make much difference whether you're on a bicycle or in a Lear jet—a race car is a race car. The Indy cars are a little heavy on the slow corners, but once you get going, it's fine. It's obvious that the cars were designed for the big speedways, but they are very good on a road course."

Willy T. Ribbs, who will become the first black driver in a sanctioned Indy car race, was in 17th position after a fast lap of 85.234 m.p.h., but he did not seem overly concerned.

"We were way off balance," he said. "We're trying to get it right

Al Unser Jr. had the pole with Fittipaldi second and Mears third. For a record third year in a row, Al Unser Jr, won, a race that was once the property of the Andretti family but has now become an Unser stronghold. Since starting his amazing streak in 1988, the 28 year old has led 248 of 285 laps over the seaside street circuit and amassed $359,228 in official prize money. His take was $143,908. He led 91 of 95 laps, but late in the race, had to hold off the challenge of Penske teammates Emerson Fittipaldi and Danny Sullivan for the victory. On lap 2, Fittipaldi and Sullivan banged wheels, causing Sullivan to spin and collect Michael Andretti. Both Sullivan and Andretti recovered and charged up through the field as the race went progressed. With Unser Jr. holding a 10 second lead, a caution came out on lap 66 which bunched the field and erased Unser's advantage. All of the leaders pitted, and when the green flag came back out on lap 70, Fittipaldi was able to close up behind Unser. Fittipaldi got within two car lengths, but Unser held on for the

victory. After the early altercation, Danny Sullivan and Michael Andretti finished third and fourth.

Paul Tracy won the American Racing Series event with Mark Smith second and Tommy Byrne third. Mark Dismore won the Toyota Atlantic Race followed by Brian Till and Sandy Dells. Dorsey Schroeder won the IMSA GTO / GTU race with Pete Halsmer in second and Jeremy Dale in third. The Toyota Celebrity Race was won by Bobby Rahal who was also the first professional followed by the first amateur, Stephen Baldwin with Paul Page in third.

Little Al still owns the Long Beach Grand Prix

LONG BEACH (AP) — This time, Al Unser Jr. left no room for doubt.

Last year, Unser stirred controversy and drew a chorus of derision from the fans by shunting aside Mario Andretti on the way to winning the Toyota Grand Prix of Long Beach.

But on Sunday, he took no chances, staying in front for all but four laps on the way to a record-setting third straight Long Beach victory.

"All I want to say is that we did it right this time," Unser said after holding off a late charge by tenacious Emerson Fittipaldi.

Unser, who led for 91 of 95 laps on the 1.67-mile, 11-turn downtown street circuit, said he didn't feel the victory came easily.

"It was very competitive," the freckle-faced 28-year-old said. "Emerson had me worried big time there at the end.

"He caught me and really got on me. But today I had straightaway speed that we didn't have earlier this week."

Fittipaldi, the defending CART

AL UNSER JR.
Three straight in Long Beach

He drove his Chevrolet-powered

second for most of the final 20 laps.

But Unser used lapped traffic and an apparent horsepower advantage on the straightaways to pull steadily away over the last three laps.

Sullivan, one of Fittipaldi's Penske Racing teammates, fell back and barely held off Michael Andretti for third, with Mario Andretti — Michael's father — the last car on the lead lap in fifth.

The finish was amazing considering that Fittipaldi, Sullivan and Michael Andretti, the 1986 Long Beach winner, were involved in a dangerous incident in the first turn of the second lap.

Sullivan tried to pass Fittipaldi on the inside on the right hand turn and the Brazilian shut the door on his teammate. The cars banged together and Sullivan spun to a brief stop.

Andretti, coming up quickly, hit the brakes and avoided hitting anyone, but killed his engine and fell to last in the 25-car field.

He got restarted and began a great charge, moving up to 17th by

(See the complete race results on the attached DVD)

Raul Boesel

Scott Brayton

Teo Fabi

Emerson Fittipaldi

Scott Goodyear

Al Unser Jr.

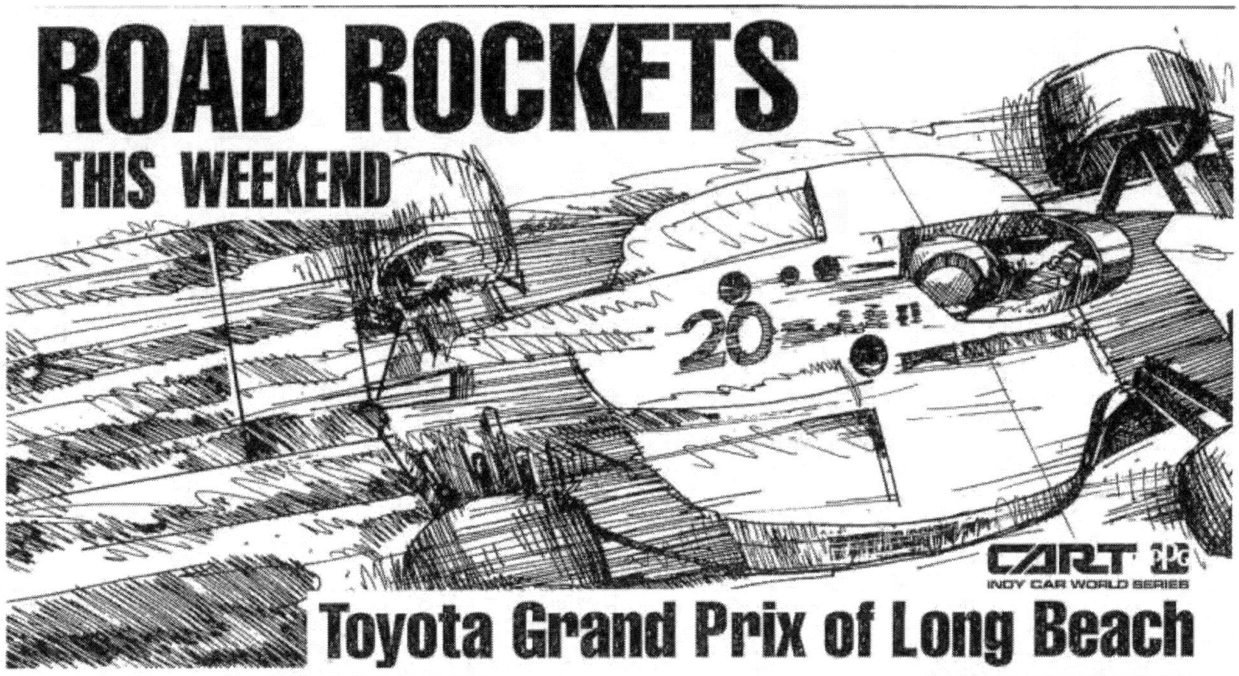

TICKET HOTLINE: (213) 436-9953 or (Outside 213, 714 area codes), call Toll-Free: 1-800-752-9524

for information or ticket purchase with a credit card; or come to our ticket office at 180 Pacific Avenue, Long Beach

Tickets Also Available at all Southern California TICKETMASTER locations.

The 16th Annual Toyota Grand Prix Returns to Long Beach

Featuring:

- CART/PPG Indy Cars
- Toyota Pro/Celebrity
- Sisapa/Toyota Atlantic Race
- American Racing Series
- Chevron GTO/USA
- Toyota One Lap of America

15th Motorsports Expo features latest in automotive displays and products

In addition to the action on the track at the Toyota Grand Prix of Long Beach, there are things to do away from the track—at the Motorsports Expo at the Long Beach Convention Center.

The show, in its 15th year, will feature local, national and international corporations exhibiting their latest products.

The hundreds of cars on display will range from old classics to prototype models. There will be the latest Fords, Chevrolets, Nissans, as well as exotic and specialty cars.

Other displays will feature such race sponsors as Toyota, Bosch, Chevron, Marlboro, Goodyear, Snap-on Tool, American Racing Equipment and First Interstate Bank. They will be joined by hundreds of manufacturers and distributors who will also be represented.

The Expo will open from 4 to 8 p.m. on Thursday. Hours will be 8 a.m. to 6 p.m. during race weekend. Admission is free to race ticket holders.

Michael Andretti

Roberto Guererro

Danny Sullivan

Bobby Rahal

American Racing Series provides drivers steppingstone to Indy cars

If the 1990 American Racing Series (ARS) season is anything like last year, it will be another season of excitement, right up to the final lap of the season.

The Long Beach ARS race is scheduled to begin at 9:45 a.m. Sunday, and will go 37 laps around the racecourse.

ARS race at the Toyota Grand Prix of Long Beach last year—along with the $18,000 first prize—and Groff are the favorites to win the series this year. Their main competition is expected to come from Gary Rubio, David Kudrave, Johnny O'Connell, Ted Prappas, Paul Tracy, Steve Shelton and P.J.

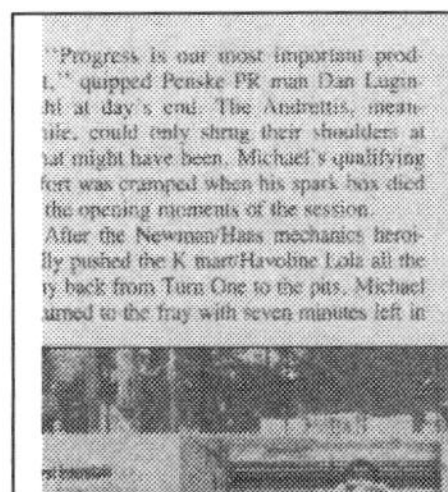
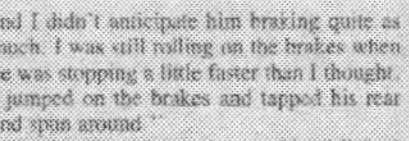

"Progress is our most important prod-t," quipped Penske PR man Dan Lugin-ht at day's end. The Andrettis, mean-ile, could only shrug their shoulders at at might have been. Michael's qualifying fort was cramped when his spark box died the opening moments of the session.

After the Newman/Haas mechanics heroi-lly pushed the K mart/Havoline Lola all the y back from Turn One to the pits, Michael urned to the fray with seven minutes left in

and I didn't anticipate him braking quite as much. I was still rolling on the brakes when he was stopping a little faster than I thought. I jumped on the brakes and tapped his rear and spun around."

An innocent bystander to this, Michael Andretti locked his brakes to avoid center-punching the stationary Sullivan. In doing so, he did a half-spin and stalled. By the time it was all sorted out, Al. Unser and Mario Andretti had pulled to quick advan-

Watch the video on our youtube channel - www.youtube.com/@racinghistoryproject2607 and read the Autoweek article on the attached DVD

1991

The Toyota Grand Prix of Long Beach, round two of the CART Championship, was held on April 14, 1991 in front of a crowd of 87,500 on race day, 204,000 for the weekend, a new record.

A busy week of events was planned, starting Tuesday with the Charity Golf Tournament at the U.S. Naval Golf Course in Los Alamitos, Wednesday had the Charity Ball at the Hyatt Regency at 6:30 PM. Thursday Long Beach City College held a mini grand prix for go karts beginning at 1:30 PM and the Indy Car Garage Preview, set for 4:00 PM. The Concours d'Promenade and Motorsports Expo was open all weekend in the Convention Center.

Friday was qualifying and practice for all groups. Saturday brought the Toyota Celebrity Race at 1:00 PM the the Exxon IMSA GTO / GTU race at 4:15 PM. Sunday had the Indy Lights race at 10:30 AM, the featured CART race at 1:00 PM and the Toyota Atlantic Race at 4:00 PM.

Michael Andretti wins pole position

Unser Jr. in Grand Prix's first row

By Nancy Miller Lewis
STAFF WRITER

Michael Andretti did something Saturday for which he's become quite famous.

He stole the pole.

In doing so, Andretti spoiled Al Unser Jr.'s try for a third-straight pole for Sunday's Long Beach Grand Prix.

GRID LOCK

The top-10 starting grid for today's 17th Grand Prix of Long Beach:

- **Row 1:** Michael Andretti (pole), Al Unser Jr.
- **Row 2:** Emerson Fittipaldi, Rick Mears.
- **Row 3:** Bobby Rahal, Scott

Michael Andretti had the pole but Al Unser Jr. won for the fourth time on a warm, sunny afternoon, Bobby Rahal was second and Eddie Cheever third. Michael beat Al into turn one at the start but the threat from Andretti and Fittipaldi vaporized when they had a pit lane collision on lap 69, not that either could have beaten Unser who led all but two laps. There was only one caution flag, when John Andretti crashed when his rear suspension broke on lap 68. That plus the pit lane crash kept the caution flag out until lap 79. It wasn't needed when the pace car, driven by Johnny Rutherford, apparently hit a patch of oil from Fittipaldi's damaged car and ran into a tire wall. Unser made $133,136 for the win.

Al Unser Jr. takes 4th-straight Toyota Long Beach Grand Prix

LONG BEACH, Calif. (AP) — The domination was nearly complete Sunday for Al Unser Jr. as he overpowered the rest of the Indy-car field and the downtown street circuit to win his fourth straight Toyota Grand Prix of Long Beach.

The second-generation driver trailed pole-winner Michael Andretti twice during the 95-lap, 158.65-mile event — the first lap and when he pitted the first time on lap 37. Otherwise, it was almost no contest for Unser, who earned his 16th CART victory.

The defending PPG Cup champion

Associated Press Photo

Michael Andretti (2) leads Al Unser Jr. (1) into Turn One at the start of the Long Beach Grand Prix.

Donny Osmond won the Toyota Celebrity Race and the amateur category with Mark Wallengren in second and Craig T. Nelson in third. The first professional, Parnelli Jones, was fifth with Dan Gurney in sixth. Steve Millen won the IMSA GTO / GTU event with Roby Gordon second and Les Lindley in third.. Eric Bachelart won the Indy Lights Race with Robbie Buhl second and Robbie Groff third. Jimmy Vasser won the Toyota Atlantic event followed by Paul Radisich and Jovy Marcelo.

(See the complete race results on the attached DVD)

Hiro Matsushita

Randy Lewis

Bobby Rahal

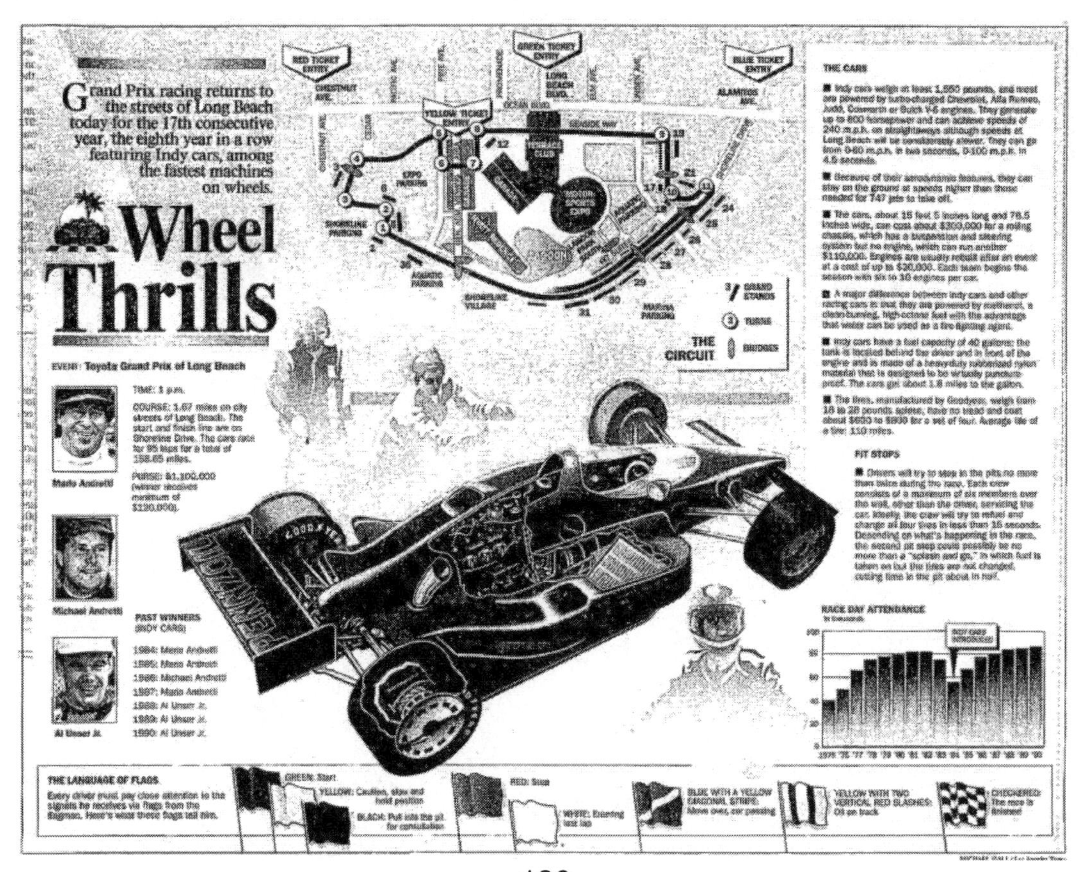

Grand Prix racing returns to the streets of Long Beach today for the 17th consecutive year, the eighth year in a row featuring Indy cars, among the fastest machines on wheels.

Wheel Thrills

EVENT: Toyota Grand Prix of Long Beach

TIME: 1 p.m.

COURSE: 1.67 miles on city streets of Long Beach. The start and finish line are on Shoreline Drive. The cars race for 95 laps for a total of 158.65 miles.

PURSE: $1,100,000 (winner receives minimum of $120,000).

Mario Andretti

Michael Andretti

Al Unser Jr.

PAST WINNERS (INDY CARS)

1984: Mario Andretti
1985: Mario Andretti
1986: Michael Andretti
1987: Mario Andretti
1988: Al Unser Jr.
1989: Al Unser Jr.
1990: Al Unser Jr.

THE CIRCUIT

RED TICKET ENTRY
GREEN TICKET ENTRY
BLUE TICKET ENTRY
YELLOW TICKET ENTRY
OCEAN BLVD.
SEASIDE WAY
TERRACE CLUB
EXPO PARKING
SHORELINE PARKING
AQUATIC PARKING
SHORELINE VILLAGE
MARINA PARKING
GRAND STANDS
TURNS
BRIDGES

THE CARS

Indy cars weigh at least 1,550 pounds, and most are powered by turbo-charged Chevrolet, Alfa Romeo, Judd, Cosworth or Buick V-6 engines. They generate up to 800 horsepower and can achieve speeds of 240 m.p.h. on straightaways although speeds at Long Beach will be considerably slower. They can go from 0-60 m.p.h. in two seconds, 0-100 m.p.h. in 4.5 seconds.

Because of their aerodynamic features, they can stay on the ground at speeds higher than those needed for 747 jets to take off.

The cars, about 15 feet 5 inches long and 78.5 inches wide, can cost about $300,000 for a rolling chassis, which has a suspension and steering system but no engine, which can run another $110,000. Engines are usually rebuilt after an event at a cost of up to $20,000. Each team begins the season with six to 10 engines per car.

A major difference between Indy cars and other racing cars is that they are powered by methanol, a clean-burning, high-octane fuel with the advantage that water can be used as a fire-fighting agent.

Indy cars have a fuel capacity of 40 gallons; the tank is located behind the driver and in front of the engine and is made of a heavy-duty rubberized nylon material that is designed to be virtually puncture-proof. The cars get about 1.8 miles to the gallon.

The tires, manufactured by Goodyear, weigh from 18 to 28 pounds apiece, have no tread and cost about $600 to $800 for a set of four. Average life of a tire: 110 miles.

PIT STOPS

Drivers will try to stop in the pits no more than twice during the race. Each crew consists of a maximum of six members over the wall, other than the driver, servicing the car. Ideally, the crew will try to refuel and change all four tires in less than 15 seconds. Depending on what's happening in the race, the second pit stop could possibly be no more than a "splash and go," in which fuel is taken on but the tires are not changed, cutting time in the pit about in half.

RACE DAY ATTENDANCE
In thousands

INDY CARS INTRODUCED

THE LANGUAGE OF FLAGS

Every driver must pay close attention to the signals he receives via flags from the flagman. Here's what these flags tell him.

GREEN: Start
YELLOW: Caution, slow and hold position
BLACK: Pull into the pit for consultation
RED: Stop
WHITE: Entering last lap
BLUE WITH A YELLOW DIAGONAL STRIPE: Move over, car passing
YELLOW WITH TWO VERTICAL RED SLASHES: Oil on track
CHECKERED: The race is finished

Paul Tracy

Driven Out

Homeless Blame Grand Prix for Eviction at Camp Site

By ROXANA KOPETMAN
TIMES STAFF WRITER

LONG BEACH—As the city spruces up for the Grand Prix, workers are gearing up to raze a homeless encampment of makeshift tents and scattered debris.

About 40 people have been told to pick up their meager belongings and leave the downtown encampment that is set up under the trestles of an old railway near the Broadway exit from the Long Beach (710) Freeway. On Sunday, workers plan to move in and remove anything left behind.

By the time thousands of tourists pour into the city for next week's Toyota Grand Prix, the area visible from the freeway will be well groomed. The three days of racing are expectd to draw 200,000 visitors. It opens Friday, April 12, with practice and qualifying laps; a celebrity race will be held April 13, and several races, including the 95-lap main event, will take place on April 14.

Ernie Spaulding, 62, right, and a man called Steve have lived under the trestle off and on for years. They say they'll come back after the Grand Prix is over.

"When you invite friends over to a party, you clean your house, don't you? So does the city," explained Sheila Pagnani, the homeless-services coordinator for Long Beach. "The city is having a big celebration. And they want to look good."

Signs warning of the cleanup have been posted in the camp, the largest in Long Beach, since last week. Some of the men and women living there said they plan to be out by Sunday. One said he wasn't budging. None knew where they would go.

"Do you know where you are going?" Mary, who has been homeless for two camp for nearly two years, shook her head.

"Me neither," Mary said.

On Tuesday, a handful of homeless people arrived unexpectedly at the City Council meeting to complain that the city is more interested in its tourists than its homeless. City Manager James C. Hankla told the City Council that the timing of the cleanup and the Grand Prix is coincidental, and officials pledged to try to find shelter for the homeless.

But some of the homeless people living at the camp noted that the city also cleaned the area at about the same Prix. They don't want the tourists to see the homeless," said Mollum, 39, who lives in the camp with her husband, John, and their dog, Daisy.

The Mollums said a city worker told them that it would be safe to return to the site the week after the Grand Prix.

Marshall Blesofsky, an activist with Long Beach Area Citizens Involved, said he saw city workers dump the homeless' belongings during a cleanup last spring, before the Grand Prix. "It was heartless," he said.

But city officials said the area poses health and safety hazards and is subject

The IMSA GTO / GTU Rce

John, Michael, Jeff and Mario Andretti

Arie Luyendyk

Al Unser Jr.

Rick Mears

COMPETITION

CART Toyota Grand Prix of Long Beach, Calif.

Grand slam

But that term means one thing to Al Jr., something else to Michael and Emmo

By David Phillips

You'd think that after three years of throwing gopher balls to Al Unser Jr. at Long Beach, the Indycar competition would have done its best to pitch a shutout in 1991. Not hardly.

First the battery of Michael Andretti and Emerson Fittipaldi spotted Al an early lead, then took themselves out of any chance for a late-inning rally with a suicide-squeeze play that left one K mart/Havoline Lola and one Marlboro Penske in shambles on pit lane and two Galles/Kraco Lolas circling the

Read the Autoweek article on the attached DVD and watch the video on our youtube channel - www.youtube.com/@racinghistoryproject2607

1992

The 1992 Toyota Grand Prix of Long Beach, race three of the CART season, was held on Aprll 12, 1992. This year featured some course changes, adding a straight and removing three corners and shortening the course to 1.59 miles. There are now five big Diamond Vision color video screens, up from three last year, at prime locations around the course

Tony Bettenhausen won the PPG Charity Golf Tournament on Tuesday. All week, The Automobile In Art Gallery on Pine Avenue will display photographs, paintings and posters of the Grand Prix. Friday was practice and qualifying with the Toyota Celebrity Race at 1:00 PM on Saturday and the Trans Am Race at 4:15 PM.

Sunday featured the Indy Lights Race at 10:10 AM, the CART Race at 1:00 PM and the Toyota Atlantic Race at 3:30 PM. The race was televised on ABC with commentators Bobby Unser, Sam Posey and Jack Arute. Paul Page was the announcer.

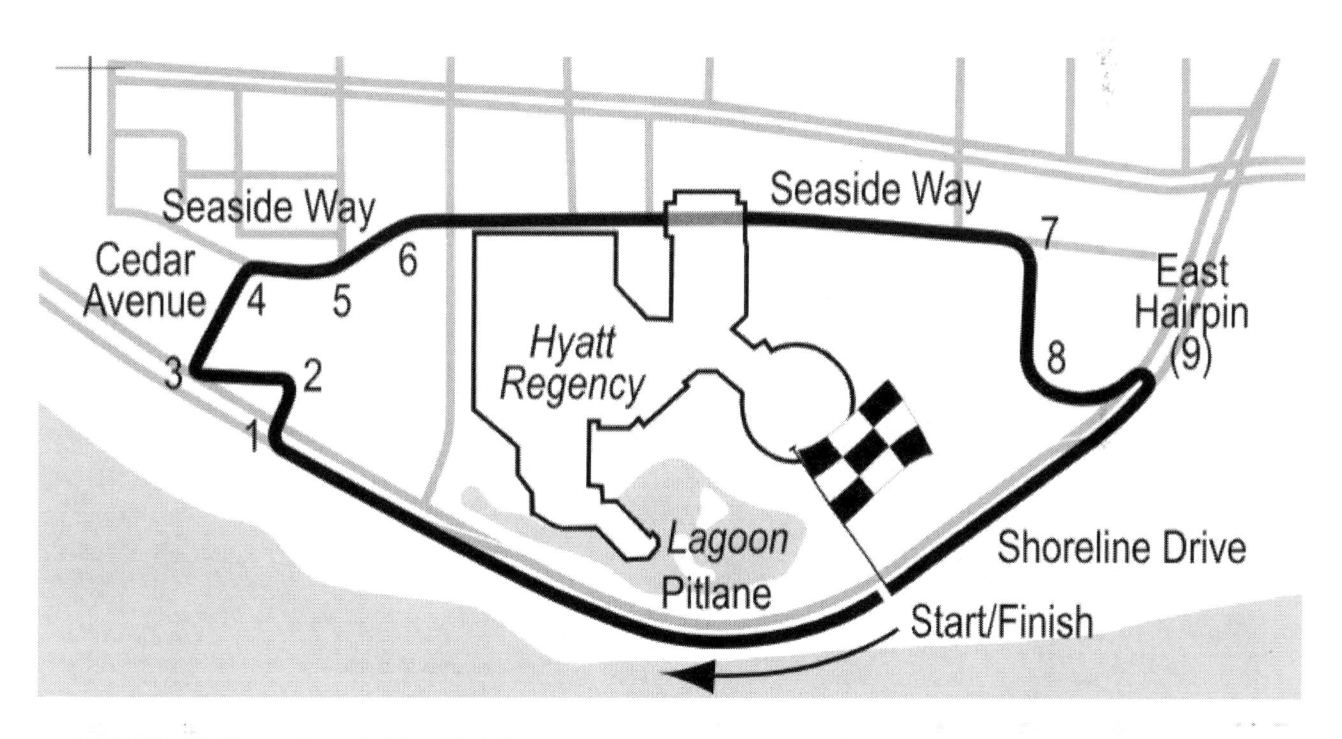

MOTOR RACING GRAND PRIX OF LONG BEACH

Michael Andretti Sets Record, Wins Pole

By SHAV GLICK
TIMES STAFF WRITER

LONG BEACH—For the eight years that the Toyota Grand Prix of Long Beach has been run, the winner has been either an Andretti

the first road course test for the new car, designed to accept the smaller Ford engine, after its oval course debut Sunday in Phoenix.

"It's a good thing we did the test," Michael Andretti said. "We learned a few things and worked

been better than Michael's. The car felt good from the moment I went out for the qualifying session. It was shortly after I had the fast lap that I had the mechanical problem. It was just an engine failure."

The Truesports car, like Michael

winner of last week's race in Phoenix.

As did Pruett, Sullivan was striving for an even quicker lap after his fast one when he lost control and clipped a wall. This turned him straight into a tire wall.

Michael Andretti had the pole but Danny Sullivan won, followed by Bobby Rahal and Emerson Fittipaldi. For 103 laps Al Unser Jr. led then as teammates Al Unser Jr. and Danny Sullivan were positioned one and two and clipping off the laps toward a storybook finish for the Galles Kraco Racing team. *"I was thinking about Little Al,"* Sullivan said. *"We were running for so long like that I literally thought, 'Well, let's just not make any mistakes and have a nice 1-2 finish."* Sullivan didn't get his wish. What he got instead was the victory but it came at a cost. With three laps to go, Unser Jr. swung wide into turn six at the end of the Seaside Way straightaway and Sullivan dived deep under his teammate in a daring passing attempt. Little Al, apparently unaware of Sullivan's position, swung back into his teammate's path. The two collided, Sullivan's nose and wing touching Unser Jr.'s right rear tire. Sullivan hung on to win with Unser Jr. finishing fourth.

Sullivan bumps off Unser to end streak

By Todd Harmonson
STAFF WRITER

When drivers come to the Long Beach Grand Prix, they know they have to knock off

was either first or second for the final 61 laps.

Unser, a master of the sharp curves featured on the 1.59-mile, eight-turn course, was leading a tight group of four in lap 102 of 105.

ALAN LESSIG/The Sun

Danny Sullivan waves to the crowd Sunday after winning the Long Beach Grand Prix.

(See the complete race results on the attached DVD)

P.J. Jones won the Toyota Celebrity Race and the professional class followed by Rod Millen and amateur class winner Joe Amato, Robby Gordon won the Trans Am Race, followed by Paul Gentilozzi and Ron Fellows. Franck Freon won in Indy Lights with Robbie Groff second and Robbie Buhl third. The Toyota Atlantic Race was won by Mark Dismore with Jamie Galles second and Chris Smith third.

Ted Prappas

Michael Andretti

Eddie Cheever

Fabrizio Barbazza

Security Keeps Ahead of Pack at Grand Prix

John Jankowski does not have the name recognition of Al Unser Jr. or Mario Andretti or Danny Sullivan or Rick Mears, but his contribution to next weekend's Long Beach Grand Prix is just about as great.

As the race's chief of security, Jankowski is responsible for the safety and comfort of more than 200,000 people. And in the eight years since Jankowski took over, spectators have yet to suffer major injury, he said.

"We get fights and drunks and forged credentials," Jankowski said, "and the occasional stolen wallet or purse. But even that doesn't happen very often because there's no place to escape."

On Wednesday at 9 a.m., Jankowski's security team will begin shutting down westbound Shoreline Drive. At 6 p.m., eastbound Shoreline Drive will be closed. Shoreline Drive will remain closed to regular traffic until the afternoon of April 13.

On race days, Jankowski sequesters himself in the middle of the crowds, cars and racers in security central, where he hunkers down for 16 hours in front of blinking consoles and waits for walkie-talkie reports from his team of nearly 900 volunteer and professional security officers.

Despite being a big Grand Prix fan, Jankowski has yet to see a race. "Only if I happen to pass it on a television console," he said, "in between putting out one brush fire after another."

Jankowski, 55, a 25-year veteran of the Long Beach Police Department, loves the racing but never considered making racing a career. At 6-foot-4 and 275 pounds, "I'm way too big to fit in one of those things," he said. "But I love it, the speed, the competition, man versus machine. It's a twofold battle. Man must conquer the machine. Then he must work with the machine to conquer the competition."

Jankowski sees a similar challenge in his line of work. "Security is like a chess game. You have to stay five or six moves ahead or the whole operation will go down on you," he said. "When the challenge is over, we evaluate it by who got hurt. If nobody got hurt, security has done its job. Then it's time to start planning for the next year."

This year, the 100-lap main event features four-time Grand Prix winner and defending champion Al Unser Jr., along with world-class drivers Mario and Michael Andretti, Emerson Fittipaldi, Sullivan, Mears and Bobby Rahal.

Qualifying rounds begin at 8:15 a.m. Friday and continue until 5:30 p.m. On Saturday, qualifying resumes at 8:15 a.m. At 1 p.m. Saturday, the pro/celebrity race begins, followed by the Tide Trans-Am Tour at 4:15 p.m., featuring Chevy Camaros and Ford Mustangs.

On April 12 at 10:30 a.m., aspiring drivers will compete in the Texaco System 3 Challenge race; the Grand Prix main event begins at 1 p.m. At 3:30 p.m., the Toyota Atlantic Championship features open-wheel cars with Toyota engines.

The total purse is more than $1 million, with the Grand Prix champion winning at least $120,000, according to race organizers.

LORI SHEPLER / Los Angeles Times

Al Unser Jr. holds up four fingers after his unprecedented fourth straight victory last year in the Long Beach Grand Prix. This year the course is faster.

□

If you go:

From Los Angeles and Orange counties, take the San Diego (405) Freeway to the Long Beach (710) Freeway, then south to its southern end and the race area in downtown Long Beach.

To avoid traffic and congestion, use Long Beach Transit (phone (310) 591-2301) or the Metro Blue Line (phone (213) 626-4455).

Private and public parking lots will be open around the race area for $5 and up. Arrive early to assure adequate parking.

Three-day passes range from $30 general admission to $85 for reserved seats. One-day tickets range from $18 to $40. Call (310) 436-9953 to order.

For more information, call (310) 437-0341.

—SUSAN PATERNO

PACE CAR
PACE CAR
UNLIKE THAT OTHER LONG BEACH LANDMARK, THIS ONE REALLY FLIES.
TOYOTA GRAND PRIX
For the eighteenth year Toyota will help some of the world's fastest cars take off during the Toyota Grand Prix of Long Beach. The 200 horses of turbo-charged flying power of the Celica All-Trac Turbo will keep even "Little Al" Unser in line while pacing all three days of the Grand Prix. So check your local listings for the April 12th telecast on ABC. And watch the All-Trac Turbo, Toyota's most aggressive and sporty Celica, really fly around Long Beach.
"I love what you do for me."
TOYOTA

Gregor Foitek

Brian Till

Rick Mears

Scott Pruett

WALLY SKALIJ/STAFF PHOTOGRAPHER

Joe Amato, left, congratulates P.J. Jones after winning Saturday. The Rolling Hills driver beat a field of television and music stars and other drivers in the pro/celebrity race.

Stars learn a few tricks from Jones

Driver captures pro/celebrity race

By Todd Harmonson
STAFF WRITER

When inexperienced drivers take the wheels of high-speed cars, there are two good places to be: Well in front of them or off the track altogether.

P.J. Jones is young and, yes, a bit cocky, so he decided to show some lead-footed celebrities how to drive.

The Rolling Hills Estates resident beat a field of TV and music stars and other drivers Saturday in the pro/celebrity race in conjunction with today's Long Beach Grand Prix.

"I got a break early and was able to slip through a couple of the celebrities," said Jones, 22, a graduate of Miraleste High.

"That helped me put some distance between me and the other pro drivers."

The race was to benefit Racing For Kids, a charitable organization that helps local children's hospitals.

But it wasn't all fun for the Jones family Saturday.

Jones' Indianapolis 500-winning father, Parnelli, was mired in the pack when his car had clutch trouble.

"I thought I'd be a good sport about it and stay out for the whole race," the elder Jones said. "It took some of the fun out of it for me."

Winning, P.J. admitted, is fun, but he would've like to edge his father at the finish line.

"I don't know what happened to him," said P.J. Jones, who drives a GTP car for Dan Gurney's racing team. "I wanted us to go side-by-side for 10 laps."

Joe Amato, a world champion in top fuel dragsters, was classified as a celebrity and finished third in the race behind pros P.J. Jones and Rod Millen.

"They made me a celebrity when they saw how much of an amateur I made of myself at driving school," said Amato.

COMPETITION

CART Toyota Grand Prix of Long Beach, Calif.

Grand slam

But that term means one thing to Al Jr., something else to Michael and Emmo

By David Phillips

You'd think that after three years of throwing gopher balls to Al Unser Jr. at Long Beach, the Indycar competition would have done its best to pitch a shutout in 1991. Not hardly.

First the battery of Michael Andretti and Emerson Fittipaldi spotted Al an early lead, then took themselves out of any chance for a late-inning rally with a suicide-squeeze play that left one K mart/Havoline Lola and one Marlboro Penske in shambles on the lane and two Galles/Kraco Lolas circling the [illegible]

Weekend support casts prove highly entertaining

Roush romps in Trans-Am; tire woes tag ARCA winner

Robby Gordon got Ford off on the right foot with his impressive victory in the SCCA Tide Trans-Am Tour season opener at Long Beach. Driving a Whistler Mustang for Roush Racing, Gordon made the most of his Trans-Am debut, starting from the pole and holding off a late-race challenge.

With 10 laps re-

Read the Autoweek articles on the attached DVD

Watch the video on our youtube channel - www.youtube.com/@racinghistoryproject2607

MOTOR RACING / SHAV GLICK

They'll Mix It Up With Celebrities at Long Beach

The Unsers, Andrettis, Mearses and Rahals—headliners in this weekend's Indy car race through the streets of Long Beach—are far from being the only racing entertainment planned for the Toyota Grand Prix, which begins Friday.

Three professional races and a pro-celebrity event are programmed around the $1-million Indy car race.

After a day of practice and qualifying for all classes Friday, there will be main events Saturday for the pro-celebrities and Sports Car Club of America Trans-Am sedans, along with the final qualifying session for the Indy cars. The pro-celebrity race, featuring entertainment and racing personalities, will be 10 laps as drivers such as Craig T. Nelson of "Coach," Tim Allen of "Home Improvement," Ian Ziering of "Beverly Hills, 90210," Richard Grieco of "21 Jump Street" and Larry Drake and Amanda Donohoe of "L.A. Law" try to keep from being run down by professional drivers Parnelli and P.J. Jones, Joe Amato and Rod Millen. The pros will start 30 seconds behind the others.

APRIL 10 - 12: A 200 MPH BEACH PARTY!

UNSER VS. ANDRETTI

Last year Al Unser Jr's fourth straight win tied Mario Andretti's Toyota Grand Prix of Long Beach record.

Now the strongest Andretti threat yet to "Little Al's" lock on Long Beach has emerged. Mario's son Michael Andretti won a record eight races and the CART PPG Cup in 1991. His next task is to topple Al Jr. from his Long Beach throne.

But Rahal, Sullivan, Mears and Fittipaldi will also be looking for a win at Long Beach.

A full weekend of action also includes the Toyota Pro/Celebrity Race, the Texaco System 3 Challenge for Firestone Indy Lights, the Toyota Atlantic Championship and the Tide Trans-Am Tour.

CART PPG
INDY CAR WORLD SERIES

FOR TICKETS CALL: (310) 436-9953 or 1-800-752-9524

TOYOTA GRAND PRIX
TICKET OFFICE,
180 Pacific Avenue,
Long Beach, Calif., or
TICKETMASTER.

Presented for charity by the
Los Angeles Times

Chris Pook and Miss Toyota Grand Prix at the Charity Ball

1993

The 1993 Toyota Grand Prix of Long Beach, race three of the CART season, was held on Aprll 18, 1993 in front of a crowd of 85,000. Beginning on Thursday was the Concours d'Promenade in downtown Long Beach featuring the Tide Trans Am Pit Stop Competition, Miss Toyota Grand Prix and her court and a world record attempt at line dancing. Friday was devoted to practice and qualifying for all.

Saturday featured the Toyota Celebrity Race at 1:20 PM , more practice and qualifying and the Tide Trans Am Race at 4:30 PM. Sunday morning brought a first; the Edison Electric Car Grand Prix at 9:35 AM, the Firestone Indy Lights Race at 10:30 AM with the CART race at 1:00 PM followed by the Toyota Atlantic race at 3:30 PM.

The race was broadcast on ABC with commentators Bobby Unser, Sam Posey,and Jack Arute. The announcer was Paul Page.

Mansell breaks track record

■ Englishmen eclipses own mark to win pole for today's

107.659 mph on the 1.59-mile, nine-turn downtown street circuit. Mansell went into the final qualifying session hoping to take it

followed by Friday's runner-up Scott Goodyear of Canada at 107.529.

The biggest jump of the second

event with a speed of 117.017.

Bodine and Earnhardt were the only drivers in the 34-car field to crack the 117 mph barrier.

Paul Tracy won, his first of four wins, and collected $87,500 for the win. Bobby Rahal was second and polesitter Nigel Mansell in third, running without second gear. Tracy had two flat tires but was able benefit from all the yellows, save fuel and made up lost time by skipping a last pit stop. Mario Andretti and Stefan Johansson crashed before the start, putting Johansson out. The race restarted on lap 5. Tracy's flat tire changes put Mansell in the lead. Al Unser Jr, attempting tp pass Mansell crashed into one of the concrete barriers, ending his race.

Crashes highlight day at Long Beach Grand Prix

By LEAH REITER

he wasn't going to take the blame.

A lot of incidents; three full course yellows and one crewman injured when he was hit by Scott Pruett in the pit lane. Robby Gordon, after an incident with Eddie Cheever, had his car repaired and while several laps down, intentionally hit Cheever in retaliation, generating a $5000 fine and probation. Danny Sullivan and Scott Goodyear both had pit

lane speeding penalties and Arie Luyendyk received a $5000 fine for unsportsmanlike conduct.

Tracy wins Long Beach GP

By Leah Reiter
Sun Assistant Sports Editor

LONG BEACH — The Long Beach Grand Prix got off to an inauspicious start Sunday when driver Stefan Johansson was knocked into the wall and out of the race by Mario Andretti before the end of the first lap.

After a bumping, grinding, roller-coaster ride, Paul Tracy emerged the winner in a record time despite finishing low on gas, overcoming two flat tires and blistering another along the way.

Even with all his problems, Tracy averaged 93.089 mph, topping last

the green flag waved on the fifth lap and traded leads with third-place finisher Nigel Mansell throughout the 105-lap race around the 1.59-mile circuit.

"I knew that the team was capable, the car was capable and, inside, I knew that I was capable of it," said Tracy, who was making only his 18th Indy-car start.

But he's been no stranger to Long Beach, having won the Indy Lights division here in 1990.

Going into Long Beach, Tracy had led in each of the first two races on the circuit this season, but had yet to finish. His best previous finishes were

lap 72 when Robby Gordon and Eddie Cheever bumped in a hairpin turn.

Mansell and several others took the opportunity to hit the pits, but Tracy's crew said he had enough fuel to finish the race.

Even with the pit stop, Mansell was just hoping he could finish the race.

He was beaten out of the pits by Mario Andretti — "It was interesting being overtaken by my teammate in the pits," said Mansell. "I've never had that happen before." — and then he slipped back with transmission problems.

Rahal wasn't even using his own

CHRIS MARTINEZ / *Associated Press*

Winning driver Paul Tracy gets one of his two flat tires on Turn 6 during the Long Beach Grand Prix Sunday.

Steve Robertson won the Indy lights Race with Franck Freon in second and Bryan Herta in third. Claude Bourbonais won the Toyota Atlantic Race with Jacques Villeneuve second and David Empringham in third. . Eddie Lawson won the Toyota Celebrity Race and the professional class with amateurs Rick Kirkham and Mark Staines in second and third. Ron Fellows win the Tide Trans Am Race followed by Bobby Archer and Dorsey Schroeder.

(See the complete race results on the attached DVD)

Dallas-Ex Dorsett Shifting Gears Into Celebrity Racing

FAST LANE

PEW! The smell was awful.

Tony Dorsett had just burned up the clutch in a brand new Geo Storm. He couldn't believe his bad luck.

"I've already torn up one car in the Long Beach (Celebrity) Grand Prix," said Dorsett, wearing a crash helmet filled with autographs collected at the event, including those of Craig T. Nelson and Leeza Gibbons. "I had a wreck during qualifying.

"I've always liked doing things fast. When I was an athlete, I ran fast. Yeah, I do things fast."

Dorsett was at Hallett Motor Racing Circuit on Tuesday practicing for this weekend's celebrity race at the Dallas Grand Prix. He will compete against former NFL quarterback Dan Pastorini, among others, with Hall of Fame running back Walter Payton expected to drive in the professional Trans-Am event. Payton crashed and burned earlier in the year and the status of his ride is in question according to event promoter Harry Farnham.

"I'm going to compete in all the celebrity races I can," said Dorsett. "I'll even drive (professionally) if I can get someone to sponsor me. I might show Payton what it's all about."

While it seems more and more former NFL greats are turning to motorsports, Dorsett probably won't be one of them. He enjoys the hoopla surrounding the celebrity races and shucking and jiving with anyone handy. His driving skills obviously need polish.

"I've had a lot of fun doing this," said Dorsett, the former Dallas Cowboys All-Pro running back. "I drove at the Long Beach Grand Prix two years ago and it was more fun than I thought it would be. I like the speed and the competition. I like being in control.

"Driving a race car takes a lot of concentration. It takes the skills of an athlete to stay focused. I just tried it for the fun and because it was something to do. I've always liked fast-moving objects and driving at high speed, but I don't have any ambition to be a race car driver per se."

Entrants in the Dallas celebrity race were supposed to undergo training in Phoenix at the Bondurant School of High Performance Driving. Dorsett couldn't make it and was put through his paces at Hallett under the watchful eye of instructor Mark Law. The celebrity race is part of a three-day event Friday through Sunday over a 1.3 mile course around Reunion Arena and downtown.

In addition to the Geo Storm celebrity race and the Trans-Am, there are five other classes, including the Oldsmobile Pro Series, the Shelby Can-Am cars and

See Tony on B-5

World staff photo by Daryl [illegible]

Former Dallas star Tony Dorsett straps on his helmet at Hallett Motor Racing Circuit.

Fellows, Lawson Win Pre-Grand Prix Races

■ Motor racing:
Canadian takes Trans-Am event, motorcycle champion the pro-celebrity contest at Long Beach.

By MIKE KUPPER
TIMES ASSISTANT SPORTS EDITOR

Ron Fellows of Toronto and motorcycle racer Eddie Lawson were winners Saturday in supporting races at the Toyota Grand Prix of Long Beach.

Pole sitter Fellows won the one-hour (54-lap) Trans-Am sedan race, the opener of the 1993 season, and Lawson was the overall winner in the pro-celebrity race.

Although he started on the pole in a Ford Mustang-Cobra, Fellows had to jump out of an Archer sandwich more than halfway through the Trans-Am to register his fourth victory in the series.

Tommy Archer of Duluth, Minn., driving a Dodge Daytona, passed Fellows for the lead at the end of the straightaway on Lap 11 and brother Bobby Archer fell in behind Fellows in another Dodge.

They ran that way until Fellows escape road after he missed a turn when his brakes locked.

"I had a great car and was able to conserve it and use it when I had to," said Fellows, who averaged 85.19 m.p.h. and won $21,000.

"It was a perfect car today. It was just up to me to bring it home."

In the pro-celebrity race, Lawson, a four-time world motorcycle champion, quickly worked his way through the celebrity field after he and the other pros had started the 10-lap race 30 seconds behind.

All drove identically prepared Toyota Celicas but Lawson was running among the top five by Lap 5, then passed both singer Peter Cetera and Rick Kirkhan of "Inside Edition" when he took the lead on the eighth lap. Kirkhan, second overall, was the celebrity winner.

"This was no different for Eddie," Kirkhan said. "He's used to going around corners on two wheels."

Lawson, who ran once last season in the Indy Lights series and wants to move into Indy car racing, agreed that the experience was familiar.

"Racing in any facet is racing," he said. "What you feel in the car is

LONG BEACH GRAND PRIX NOTES

The Pressure of Qualifying Proves a Barrier for Some

By SHAV GLICK
TIMES STAFF WRITER

The pressure of qualifying over a narrow 1.59-mile street course for the fastest field in Toyota Grand Prix of Long Beach history sent several cars into the wall Saturday.

Arie Luyendyk, Hiro Matsushita and Lyn St. James all tagged the concrete barriers, with Luyendyk's the most serious accident. When the 1990 Indianapolis 500 winner crashed into the wall at Turn 4, his Lola-Ford Cosworth suffered heavy damage to the left front suspension.

"We were still struggling and now this," he said. "I touched the inside wall and bounced into the outside wall. I needed to run the corner tight and got too close. I'm OK. I was pressing for a good lap, but when a car is reluctant to perform, it is almost impossible to carry it around this track."

Luyendyk will start today's 105-lap race in a backup car.

Matsushita crashed as he was leaving the pits, St. James as she was entering the pits.

"I am not sure what happened," Matsushita said. "Everything went so quickly. The car just snapped to the right. I knew I was going out on cold tires, so I went very easy. I was in fourth gear, low r.p.m., when it just turned."

St. James brushed the wall but

□

Nigel Mansell collected a $10,000 bonus for winning his second pole of the season. If he wins today's 167-mile race, he will receive an additional $30,000 for having swept the pole and the race.

□

The Galles Racing Team has won five consecutive Long Beach Indy car races, but its two drivers, Al Unser Jr. and Danny Sullivan, did not fare too well in qualifying.

Unser, who won four in a row between 1988 and 1991, qualified eighth, and Sullivan, last year's winner, was 12th.

"A valve spring let go and we dropped a valve," Unser said after his car quit during qualifying. "We barely even got going. The car was going good this morning when we got a pretty good idea of which way we should be going. Now we don't have time to see if we were going in the right direction."

□

Ford-powered cars won the first two Indy car races this year—Mansell in Australia and Mario Andretti at Phoenix—but the results of qualifying indicate that Chevrolet may be catching up.

In the first six positions, three are Fords and three are Chevys. The Chevrolet challengers are Penske teammates Paul Tracy, who is second, Emerson Fittipaldi,

STEVE DYKES / Los Angeles Times

Englishman Nigel Mansell broke the track record with a lap of 108.198 m.p.h. Saturday at Long Beach.

MOTOR RACING/SHAV GLICK

Long Beach Will Hum With Electricity

Mario Andretti, Nigel Mansell, Bobby Rahal and other Indy car drivers will be displaying state-of-the-art racing machines in the 19th Toyota Grand Prix of Long Beach this weekend, but what might be a preview of things to come will take place Sunday morning on the same 1.59-mile street course.

The first Edison Electric Car Grand Prix will showcase most of the world's existing electric formula race cars—capable of 110 m.p.h.—in a 20-minute demonstration race.

The race will include at least one pit stop. Instead of refueling, however, crews will swap battery packs, which requires about 20 seconds.

"This event goes a long way toward showing the public that electric vehicles are not only safe and practical, but fast and exciting, too," said Peter Quinn Hackes, co-promoter of the race with Rebecca Murray.

"Although it will be fun and competitive, the bottom line is the environment and its goal toward

Stock Car Racers to Compete at Indianapolis Speedway in '94

For the first time since 1910, the Indianapolis Motor Speedway will conduct a race in 1994 besides the Indianapolis 500.

NASCAR's Winston Cup stock cars will run the Brickyard 400 on Aug. 6, over the same 2½-mile low banked, rectangular oval where only Indy cars have traditionally raced.

"The announcement that Winston Cup cars will race here is the most important event involving Indianapolis Motor Speedway since 1945, when my grandfather [the late Tony Hulman] purchased the track," said Speedway President Tony George in a joint announcement in Indianapolis with Bill France Jr., NASCAR president. "A race in August will complement our traditional race in May."

France said: "There is probably not a race car driver in America, whatever he is driving, who doesn't want to race on the Indianapolis Motor Speedway track. Some of our cars came here for tire tests last year, and that answered a lot of questions about the feasibility of holding a stock car race here."

Unlike the Daytona 500, stock car racing's premier event, and the Indianapolis 500, which are run on Sunday, the Brickyard 400 will run on Saturday. Qualifying will be Thursday and Friday, leaving Sunday as a rain date.

Speculation about such a race has been strong since a two-day tire test last June by leading NASCAR drivers Dale Earnhardt, Darrell Waltrip, Rusty Wallace, Davey Allison, Kyle Petty, Ernie Irvan, Mark Martin, Bill Elliott and Ricky Rudd.

"I know that there's a lot of tradition surrounding the Indy 500," Earnhardt said. "It's the biggest show in racing, but it'll be a big

Herta was the series' top rookie last year after winning the Zerex Saab Pro Series in 1991.

Motor Racing Notes

MOTORCYCLES—The 23rd annual El Trial de Espana and the Schreiber Cup, premier events of the American Trials Assn. season, will be run Saturday and Sunday at Cougar Flats in the Johnson Valley off-highway vehicle area, near Lucerne Valley in San Bernardino County. The Schreiber Cup—named for **Bernie Schreiber**, the La Crescenta rider who became America's only world trials champion—is scheduled for Saturday and the El Trial de Espana for Sunday.

World superbike champion **Doug Polen** will continue his quest of his first American Motorcyclist Assn. national title this weekend at Laguna Seca Raceway, near Monterey. Polen won the season opener at Phoenix and is defending champion at Laguna Seca. National champion **Scott Russell** will also ride, but is expected to leave the series to compete for the world championship left undefended by Polen. . . . Speedway bikes will be at the Orange County Fairgrounds in Costa Mesa for their weekly Friday night program.

MIDGETS—United States Auto Club drivers will be at Ventura Raceway for a full midget and TQ doubleheader Saturday night. A large field, headed by defending

Scott Goodyear

Willie T. Ribbs

Nigel Mansell

Jeff Wood

Mark Smith

Robbie Gordon

Toyota Celebrity Race

Tracy, Rahal and Mansell,

Thursday, April 15th • 11 am-3pm
at the Long Beach Downtown Promenade
between Third Street and Ocean Blvd.
Free Admission!

Duel in the Pits!

CONCOURS D'PROMENADE PIT STOP COMPETITION

**presented by the Downtown Long Beach Associates,
the Grand Prix Association of Long Beach
and SCCA's Trans-Am Tour**

Los Angeles Times

Don't miss this event! The Tide Trans-Am pit stop competition at the downtown promenade is part of the Concours d'Promenade, the official downtown kick-off for the TOYOTA GRAND PRIX OF LONG BEACH. In addition to the pit stop competition, the afternoon excitement includes Miss Toyota Grand Prix and her court, celebrity drivers, a custom truck and car show, prizes, *a world record attempt in line dancing* and much more! **For information contact the DLBA at (310) 436-4259.**

***For Grand Prix tickets call (310) 436-9953
or visit the downtown ticket office at
430 East First Street, Long Beach***

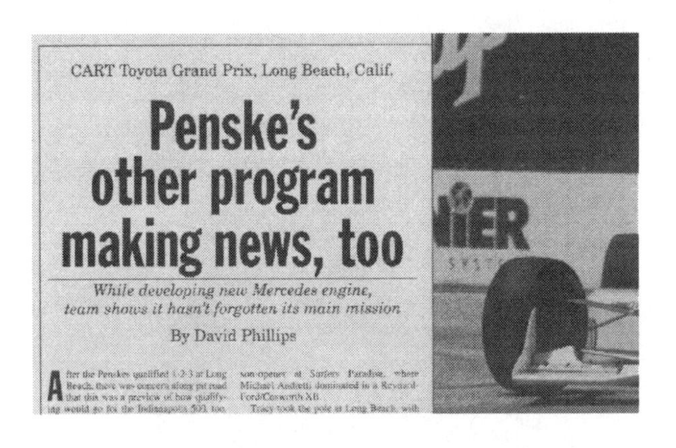

CART Toyota Grand Prix, Long Beach, Calif.

Penske's other program making news, too

While developing new Mercedes engine, team shows it hasn't forgotten its main mission

By David Phillips

After the Penskes qualified 1-2-3 at Long Beach, there was concern along pit road that this was a preview of how qualifying would go for the Indianapolis 500, too.

son-opener at Surfers Paradise, where Michael Andretti dominated in a Reynard-Ford/Cosworth XB.

Tracy took the pole at Long Beach with

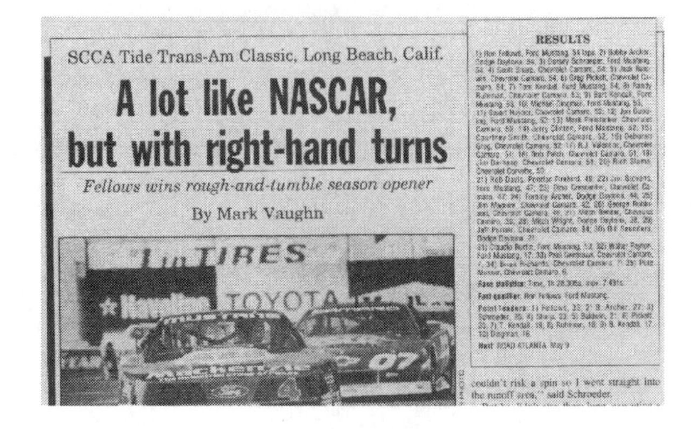

SCCA Tide Trans-Am Classic, Long Beach, Calif.

A lot like NASCAR, but with right-hand turns

Fellows wins rough-and-tumble season opener

By Mark Vaughn

RESULTS

couldn't risk a spin so I went straight into the runoff area," said Schroeder.

Read the Autoweek articles on the CART race and the Trans Am race on the attached DVD

Watch the video on our youtube channel - https://www.youtube.com/@racinghistoryproject2607

1994

The 1994 Toyota Grand Prix of Long Beach, race three of the CART season, was held on Aprll 17, 1994 in front of a crowd of 84,000. The week before, over 6000 athletes took part in the Ride, Run and Roll Grand Prix Sports Festival which included 18 competitive events ranging from dual slalom mountain bike racing to amateur and professional dualthalon competition and 5K runs.

Beginning on Thursday was the Concours d'Promenade in downtown Long Beach featuring the Firestone Indy Lights Pit Stop Competition, Power Ranger Bicycle Team, Kids Racing Association and Miss Toyota Grand Prix and her court. Friday was devoted to practice and qualifying for all.

Saturday featured the Toyota Celebrity Race at 1:20 PM, the Exide Electric Grand Prix at 2:00 PM and the Bridgestone Supercar Race at 3:00 PM. Sunday, race day, had the Firestone Indy Lights Race at 10:30 AM with the CART Race at 1:00 PM followed by the Toyota Atlantic race at 3:30 PM.

The race was live on ABC, tape delayed in the Los Angeles area. Commentators were Bobby unser, Sam Posey and Jack Arute. Paul Page was the announcer. The Grand Prix of Long Beach Association announced it was going public, raising thirteen million dollars when its stock opened at $13 per share.

Tracy grabs pole slot

Associated Press

LONG BEACH — Penske Racing teammates Paul Tracy, Al Unser Jr. and Emerson Fittipaldi all broke the track qualifying record Saturday while sweeping the top three positions for the Grand Prix of Long Beach.

But Unser, the newest Penske driver and a four-time Long Beach winner, was close behind at 108.407. Fittipaldi, who was second to Nigel Mansell in Friday's provisional qualifying, improved to 108.308 but fell to the second row of the 30-car field.

The three fast laps, all quicker than Mansell's year-old record of

Al Unser Jr won his fifth Long Beach Grand Prix under cooler than usual, cloudy conditions and overcame a stop and go penalty for speeding in the pit lane. He collected $67,500 for the win. Nigel Mansell was second, 39 seconds behind. Robbie Gordon was third. Paul Tracy got the pole but went out after 75 laps with gearbox and brake problems.

Fittipaldi briefly threatened but went out with a broken gearbox on lap 66. Mario Andretti, in his final Long Beach appearance, was fifth.

In tradition, Unser Jr. captures fifth

■ 'Little Al' overcomes a penalty for a 39.107-second victory over England's Nigel Mansell.

By PAUL OBERJUERGE
Sun Sports Editor

LONG BEACH — The Indy-car arithmetic is simple. Unser-plus-Penske equals serious trouble for everyone else.

Bobby Unser. Al Unser. And now Al Unser Jr.

"Little Al," thoroughly enjoying his first year racing with the formidable Penske team, overcame a potentially

The victory was Unser's fifth over the city streets of this seaside community and 20th in his career. It came two days before his 32nd birthday.

"This team sets the standard in Indy-car racing," Unser said. "I'm proud and grateful to be here."

A year ago, Unser was racing for the Galles team, and hating it. He feuded with teammate Danny Sullivan, questioned Galles' commitment to his car — and won only one race.

When the chance to drive for Penske came up, even with Paul Tracy and Emerson Fittipaldi already on board, "Little Al" jumped at it.

that's as good as it gets," Unser Jr. said.

For a time Sunday, a 1-2-3 finish by Team Penske seemed likely. But Tracy, the pole-sitter, had problems with his brakes and gearbox, spun out four times and went out after 75 laps. Fittipaldi smoked his gear box, as well, and gave up after 66 laps.

Unser was forewarned as well as forearmed.

"Roger told me (via headset) to be careful with the gearbox," Unser said. "We already knew it was a weakness."

He first took the lead after Tracy's first spin, then lost it to Fittipaldi when he was penalized for exceeding the 60

first stop," he said. "These machines accelerate so fast. I realized I was speeding and tried to slow down, but I got caught."

Unser moved back ahead when Fittipaldi got in trouble, and pulled away to a whopping 39.107-second victory over England's Nigel Mansell. Rising star Robby Gordon, 25, a native of Orange, was third and Raul Boesel of Brazil fourth.

Mario Andretti was a solid fifth in his final LBGP appearance, and son Michael was sixth.

The 1.59-mile temporary road course was slick and produced numerous spins, but no apparent injuries and no serious accidents.

UNSER JR.'S CAREER GLANCE

- **Birthdate:** April 19, 1962.
- **Height:** 5-10.
- **Weight:** 150.
- **Residence:** Albuquerque, N.M.
- **Years on circuit:** 13.
- **Car:** Penske-Ilmor V8-D.
- **Indy-car wins:** 20.
- **Poles:** 3.
- **Items of interest:** At

Alfonso Ribiero won the Toyota Celebrity Race and was followed by Sean Astin, both amateurs. The race marked the return of Clay Regazzoni, paralyzed after a crash in the 1980 Formula One race, now driving with hand controls. The first professional was Brian Redman in third. David Donohue won the Bridgestone Supercar Race followed by Shawn Hendricks and Peter Farrell. Steve Robertson won the Toyota Atlantic Race followed by Richie Hearn and Pedro Chaves

Ribiero Steals Scene in Pro-Celebrity Race

By MIKE KUPPER
TIMES ASSISTANT SPORTS EDITOR

Brian Redman certainly made it interesting, but could an old-time pro beat a young charger?

Not Saturday in the pro-celebrity race at the Toyota Grand Prix of Long Beach.

Luckily for Alfonso Ribiero, though, the race went only 10 laps. Two more and "The Fresh Prince of Bel Air" co-star would have

close that fast."

Peter Farrell of Manassas, Va., was third in a Mazda RX-7.

□

Redman—who won the first Grand Prix of Long Beach, a Formula 5000 race in 1975—said he hardly recognized the place.

"The course is really completely different," he said. "Parts of the main straight are the same, but that's all. You're busy racing and

ED CARREON/STAFF PHOTOGRAPHER

Ken Stabler, former Raiders quarterback, stays ahead of drag racer Cruz Pedregon in Pro/Celebrity race Saturday.

Mario Andretti

Emerson Fittipaldi

J. ALBERT DIAZ / Los Angeles Times

1974 Long Beach winner Clay Regazzoni oversees work on

A Collision With the Past

Paralyzed Regazzoni Returns as Street Racer to Site of Tragedy

By BILL PLASCHKE
TIMES STAFF WRITER

ROSAMOND, Calif.—The buzzing and squealing of a man chasing his past echo interminably across the high desert. Then suddenly, silence.

The modified Toyota Celica, alone amid the scrub brush and tumbleweeds of a scarred asphalt track, rolls to a wall.

The car stops. Driver Clay Regazzoni raises his hands. The noise has returned and this time it is him, shouting in Italian.

steering.

He needs a belt for his left leg.

He needs it to keep the useless limb from flopping on his equally useless right leg.

He can't move the leg back with his hands, because his hands do more than steer and shift.

His hands work the gas. And the brakes. And the clutch.

The rules for this once-great driver are different now.

But his will is not.

In 1980, after winning the inaugural event four

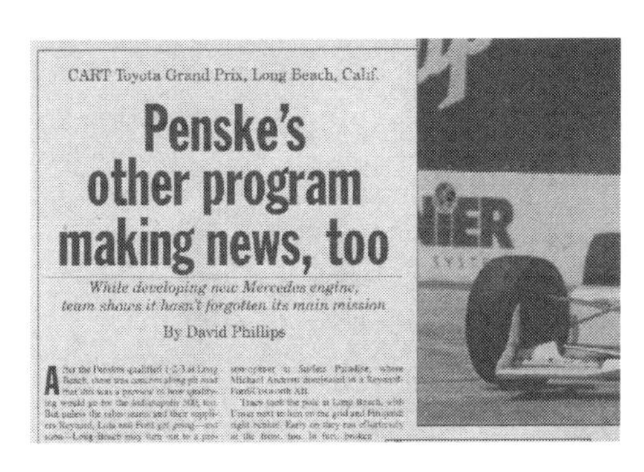

CART Toyota Grand Prix, Long Beach, Calif.

Penske's other program making news, too

While developing new Mercedes engine, team shows it hasn't forgotten its main mission

By David Phillips

Read the Autoweek article on the attached DVD and watch the video on our youtube channel - https://www.youtube.com/@racinghistoryproject2607

Arie Luyendyk

Paul Tracy

20 Years
IN THE FAST LANE!
Celebrate two decades in the fast lane with a classic Long Beach battle. 1993 winner Paul Tracy will take on PPG Indy Car Champion Nigel Mansell, the returning Michael Andretti and the other Indy Car stars!
TOYOTA GRAND PRIX OF LONG BEACH
APRIL 15 - 17
Twentieth Anniversary
TOYOTA GRAND PRIX OF LONG BEACH
FOR TICKETS, CALL NOW!
(310) 436-9953
(800) 752-9524
OR TICKETMASTER
TOYOTA GRAND PRIX TICKET OFFICE
430 EAST FIRST STREET
LONG BEACH
Los Angeles Times

Duel in the Pits!

CONCOURS D'PROMENADE PIT STOP COMPETITION

presented by the Downtown Long Beach Associates and the Grand Prix Association of Long Beach

Los Angeles Times

See the PPG-Firestone Indy Lights Championship pit stop competition, the Power Ranger Bicycle Team and the Kids Racing Association (KRA) Grand Prix at the Concours d'Promenade, the downtown kickoff of the 20TH ANNIVERSARY TOYOTA GRAND PRIX OF LONG BEACH. The afternoon excitement also includes Miss Toyota Grand Prix and her court, Grand Prix drivers, custom and race cars, prizes and a special guest appearance by MARIO ANDRETTI, Grand Marshal of the KRA Grand Prix. **For information contact the DLBA at (310) 436-4259.**

For Grand Prix tickets call (310) 436-9953 or visit the downtown ticket office at 430 East First Street, Long Beach

ASSOCIATED PRESS

Speedsters: Long Beach Grand Prix winner Al Unser Jr. is flanked on his victory lap Sunday by Englishman Nigel Mansell (second place), right, and Robby Gordon (third).

ED CARREON/STAFF PHOTOGRAPHER

Nigel Mansell leads the pack during the first day of qualifying Friday at the Long Beach Grand Prix. Mansell led the way with an average speed of 107.919 mph.

Toyota Has Kept Fast Company For The Last Twenty Years.

Presenting the 20th Toyota Grand Prix of Long Beach.

Since 1975, Toyota has been running with the fast crowd at Long Beach. And for twenty years, we've set the pace for Formula 5000, Formula I and the Indy cars. Now Supra, as the official pace car, will again get the excitement rolling for the 1994 Toyota Grand Prix of Long Beach. It's the largest race of its kind in the U.S. and the third largest motorsports event in North America. You can look forward to more great times with some of the biggest names in racing and entertainment. Fast company, fast times. Toyota does it again, April 15-17. Check your local listings for the telecast on ABC.

TOYOTA

"I love what you do for me."

CONCOURS d' PROMENADE

THURSDAY APRIL 14th • 11:00 AM - 3:00 PM

Sponsored by
DOWNTOWN LONG BEACH
ASSOCIATES AND GRAND PRIX
ASSOCIATION OF LONG BEACH

✓FREE ADMISSION
✓INDY CARS/DRIVERS
✓DOOR PRIZES
✓PIT STOP COMPETITION
✓ENTERTAINMENT
✓VENDORS

"LADIES & GENTLEMEN...
START YOUR ENGINES"

DON'T MISS THIS EVENT!

TOYOTA GRAND PRIX
OF • LONG • BEACH
APRIL 15 - 17 1994

DON'T MISS THIS EVENT! THIS WILL BE THE OFFICIAL DOWNTOWN KICK OFF FOR THE TOYOTA GRAND PRIX OF LONG BEACH. ADDITIONAL AFTERNOON ACTIVITIES TO INCLUDE...

✓ Mini Indy Grand Prix
✓ Miss Toyota Grand Prix and Her Court
✓ Kids Racing Associaton Education Day
✓ Pit Stop Competition ✓ Celebrity Drivers
✓ Monster Trucks/Dragsters/Funny Cars/Vintage Cars

FOR MORE INFORMATION CONTACT DLBA at 310/436-4259

Special Thanks to...
Grand Prix Assoc. of L.B., Committee of 300, Kids Racing Assoc.

SPECIAL ATTRACTION
Indy Preview
NIGHT
Convention Center
Thursday, April 14th
5:00 - 8:00 PM

1995

The 1995 Toyota Grand Prix of Long Beach, race three of the CART season, was held on April 9, 1995 in front of a crowd of 85,000. Total attendance for the weekend was 205,000. The highest bidder at the Charity Ball won a seat in the Toyota Celebrity Race, paying $19,000 for the ride. Thursday was the Concours d'Promenade in downtown Long Beach featuring a children's grand prix in addition to race cars and classics on display.

Friday was practice and qualifying for all. Saturday featured the Toyota Celebrity Race at 1:45 PM and the Bridgestone Supercar Race at 2:45 PM. The GT Bicycle Air Show offered daily performances outside the Expo.

Sunday, race day, had the Firestone Indy Lights Race at 10:30 AM with the CART Race at 1:00 PM followed by the Toyota Atlantic Race at 3:30 PM. The race was live on ABC, tape delayed in the Los Angeles area. Commentators were Bobby Unser, Sam Posey and Jack Arute. Paul Page was the announcer.

ED CARREON/STAFF PHOTOGRAPHER

Michael Andretti set track records for lap time and speed in winning the provisional pole for Sunday's Long Beach Grand Prix. He expects the records to fall again today.

While Michael Andretti had the pole and a new track record, Al Unser Jr. won for the sixth time, collecting $87,500. Scott Pruett was second, 23 seconds behind and Teo Fabi was third, the last car on the lead lap. Paul Tracy, starting on the front row, jumped the start, producing an immediate yellow flag. Michael Andretti led while Unser, in third, passed Tracy for second on lap 16, after which Tracy and Gil de Ferran came together putting them both out. Unser passed Andretti after a restart on lap 25 and stayed ahead. Andretti attempted a late braking pass on lap 55 and ended up in the runoff area.

Unser Jr. cruises at Long Beach Grand Prix

Dominates 105-lap event from the start

By MIKE HARRIS
AP Motorsports Writer

LONG BEACH, Calif. – Al Unser Jr. renewed his lease on the downtown street circuit at Long Beach on Sunday, turning around a disappointing season start with his sixth Toyota Grand Prix victory.

LONG BEACH GP

Unser, who has 28 career wins, won for the second straight year and the sixth in the past eight on the 1.59-mile, eight-turn temporary circuit.

It was a big win for Unser, who won eight of 16 races last season on the way to his second PPG IndyCar World Series title. The 32-year-old second-generation racer had struggled through the first three races this season, hampered by electronic gremlins, failing to finish higher than sixth.

But seeing the picturesque, waterside Long Beach circuit got him back on track as Unser, who will turn 33 on April 19, dominated the 105-lap, 166.95-mile race, averaging 91.442 mph.

AP/Reed Saxon

Starting under the yellow caution flag, Michael Andretti (6) leads the field through the first turn at the Long Beach Grand Prix

Alfonso Ribiero won the Toyota Celebrity Race and the amateur class, a second time for him, followed by Anthony Edwards and Grant Show. Rod Millen, first in the professional class, was fifth. Sean Roe won the Bridgestone Supercar Race in a Corvette, followed by Shawn Hendricls in a 300ZX and Peter Farrell in an RX-7. David Empringham won the Toyota Atlantic Race with Zeca Giaffone in second and Colin Trueman in third. Gerg Moore won the Firestone Indy Lights Race followed by Robbie Buhl and Dace DeSilva.

LONG BEACH GRAND PRIX NOTES

Ribeiro Repeats in Celebrity Run

By MIKE KUPPER
TIMES ASSISTANT SPORTS EDITOR

Alfonso Ribeiro had such a good time winning the pro-celebrity race last year at the Toyota Grand Prix of Long Beach that he relived the experience Saturday.

This time, though, he made it look easy.

Today's Long Beach Grand Prix

■ Where: Start and finish line on Shoreline Drive

■ Time: 1 p.m.

Va., who was third in a Mazda RX-7.

Roe averaged 80.377 m.p.h in the 26-lap race.

□

If Michael Andretti's performance today is as impressive as the rest of his weekend has been, he will become a two-time winner of the Long Beach Grand Prix.

He was the fastest driver in

(Read All The Race Results On the Attached DVD)

CART Toyota Grand Prix

No Change

Sixth time's the charm for Unser at Long Beach

by David Phillips

Tire: and the Patrick Racing team into the points lead of the 1995 PPG Indy Car World Series.

"This is something I couldn't even have imagined last year," said Pruett. "For the first time, I'm having fun in Indycar racing—but this is the first time I've been involved with such an experienced group of guys. We're well-funded, dedicated and I'm just proud to be a part of it."

Until lap 17 on Sunday, the weekend looked like a Newman/Haas benefit. Michael Andretti set fastest time in every

Read the Autoweek article on the attached DVD and watch the video on our youtube channel - www.youtube.com/@racinghistoryproject2607

Brian Herta

Dean Hall

Unser, Pruett and Fabi

Miss Toyota Grand Prix and Runner Ups

TOYOTA TOYOTA TOYOTA
Marlboro Marlboro Marlboro

1996

The 1996 Toyota Grand Prix of Long Beach, race four of the CART season, was held on Aprll 14, 1996 in front of a record crowd of 88,000. Thursday was the Concours d'Promenade in downtown Long Beach featuring a children's grand prix in addition to race cars and classics on display. Friday was practice and qualifying for all.

Saturday featured the Toyota Celebrity Race at 1:45 PM and the Bridgestone Supercar Race at 2:45 PM. Sunday, race day, had the Firestone Indy Lights Race at 10:30 AM with the CART Race at 1:00 PM followed by the Toyota Atlantic Race at 3:30 PM.

The race was live on ABC, tape delayed in the Los Angeles area. Commentators were Danny Sullivan and Jack Arute with Paul Page as announcer. This was the year of the CART / IRL split with CART (Championship Auto Racing Teams) headed by Roger Penske and Pat Patrick disagreeing with the IRL (Indy Racing League), headed by Tony George, president of Indianapolis Motor Speedway, who wanted more ovals and fewer road courses. Meanwhile the Grand Prix of Long Beach Association prepared to become publicly traded on NASDAQ.

Penske and George at heart of Indy-car battle

By Bill Koenig
STAFF WRITER

The fight for control of Indy-car racing is seen by many fans as a conflict between two men: Roger Penske and Tony George.

Penske, 59, heads Penske Corp., a $5 billion-a-year company that makes diesel engines and operates truck leasing firms, auto dealerships and motorsports tracks.

He's the co-founder of Championship Auto Racing Teams Inc. and one of its most successful Indy-car team owners. Penske also owns a portion of British-based Ilmor Engineering, which makes engines for CART and Formula One racing teams.

Penske's race team came to Indianapolis for 25 years, winning the 500-Mile Race 10 times. Penske has capitalized on that image success, using his drivers to help greet customers and employees of his other businesses.

Tony George, 36, is president of the Indianapolis Motor Speedway, a position he assumed in 1989. He's also head of Hulman & Co., a Terre Haute-based firm that oversees all of his family's business interests. The closely held company doesn't disclose financial data, but its holdings include television stations (such as WNDY 23 in Indianapolis and WTHI in Terre Haute) and food products such as Clabber Girl baking powder.

George began a new Indy-car circuit, the Indy Racing League, in 1994 because he felt CART wasn't providing the kind of racing the Speedway needs. The new league includes the Indianapolis 500.

Publicly, the two are reserved in their comments about each other. But privately, some people associated with the Speedway are sensitive about Penske and his position within CART.

George said he doesn't have strong personal feelings about Roger Penske.

"We've not been what you'd call the closest of friends," George said. "He's put his resources into founding Championship Auto Racing Teams and he should have all the influence he has."

Penske declined an interview request. A Penske assistant, Dan Luginbuhl, said his boss has wanted to reach a solution to the impasse. "It's gone absolutely nowhere," Luginbuhl said. "Roger does not want to engage in any further dialogue on it."

De Ferran keeps Long Beach Grand Prix pole

ASSOCIATED PRESS

Motorsports

LONG BEACH — Gil de Ferran improved just enough Saturday to hold onto the pole position for today's Toyota Grand Prix of Long Beach ... ended a string of two straight poles for tire rival Firestone.

"I expected I would have to go ...

"I could have gone faster, but I was right behind Robby when he blew the engine," Vasser said. "I had just put new tires on. I came right in. I knew they were going to throw a red (flag), and then they ...

Warmer than usual, the temperature was 94 deg. F. at race time. Gil de Ferran had the pole and led at the start. The first yellow was brought out by Andre Ribeiro on lap four. The next caution was on lap 39 when Alex Zanardi and Bobby Rahal came together in turn one, forcing Zanardi's retirement. At the same time, Robby Gordon had a pit fire. Moore and Christian Fittipaldi put themselves out with a crash on lap 49 and the resultant debris

damaged Emerson's car and he retired, Christian was later fined $5000 after a confrontation with Moore. Robby Gordon hit the wall producing another yellow flag period on lap 70. Paul Tracy received a stop and go penalty for ignoring the yellow flag and de Ferran retired with four laps to go, turning the lead over to Jimmy Vasser. Then Rahal and Herta collided, leaving Vasser to win, collecting $95,000. Parker Johnstone was second and Al Unser Jr. third.

Victory Falls in Vasser's Lap

GINA FERAZZI / Los Angeles Times

What looked to be a day of celebration turned into a crash course in disappointment for Gil de Ferran's crew. Mechanical problems on the 102nd lap dropped the leader to fifth.

■ **Grand Prix:** He registers his third victory in four races when de Ferran's car falters in 102nd lap.

By SHAV GLICK
TIMES STAFF WRITER

The record books will show that Jimmy Vasser won three of the first four races in this year's PPG Indy Car World Series, but there should be an asterisk beside the third one.

"They say sometimes you'd rather be lucky than good, and today it seemed that we were lucky," Vasser said after winning the 22nd Toyota Grand Prix of Long Beach before a record 88,000 on a picture-perfect Sunday.

For 101 laps the race belonged to Gil de Ferran, the Brazilian pole-sitter who was cruising along in Jim Hall's Reynard-Honda, easily staying ahead of a destruction derby behind him.

But the race was 105 laps.

Please see RACE, C7

Grant Show won the Toyota Celebrity Race with Jason Bateman second and Alfonso Ribeiro third. Motorcycle champ Wayne Rainey, now paralyzed and driving with hand controls. finished fourth, This year's race had no professionals, instead the field was made up of previous amateur front runners. Chase Montgomery won the Toyota Atlantic Race followed by Patrick Carpentier and Chuck West.

Celebrity Race Is One-Man Show

By MIKE KUPPER
TIMES ASSISTANT SPORTS EDITOR

Pole sitter Grant Show of "Mel-

ing those guys up with me made it fun."

This year's pro-celebrity race was like each of the previous in

"When I saw him there, I got my burners going," Montgomery said of Carpentier. "I thought it might be too late, and it almost was.

CART Toyota GP of Long Beach

Hall's Heartbreaker

Washed down by de Ferran's tears, fifth place was hard to swallow

by Sam Moses

SCCA Optima Batteries Trans-Am Classic

Properly Tired, and Happy

Young Galles listens and wins

by Mark Vaughn

Results

Read the Autoweek article on the CART race and the Trans Am on the attached DVD

Watch the video on our youtube channel - www.youtube.com/@racinghistoryproject2607

Friendlier atmosphere drew de Ferran, Zanardi to U.S.

By Nate Ryan
Sun Sports Writer

LONG BEACH — Competing with the best drivers in the world does have its disadvantages.

Along with talent, driving on the Formula One circuit takes money, power and connections. And a thick skin.

Gil de Ferran and Alex Zanardi — the two top qualifiers for today's Long Beach Grand Prix — both left the prestigious racing circuits of Europe for the world of Indy Car.

They could be happier because of the immense success they've had racing with Indy Car teams. But that's not the only reason they made the switch.

"The atmosphere is so much more relaxed here," said Zanardi, a Monte Carlo, Monaco, native who raced Formula One from 1991-94, filling in for defending world champion Mi-

"(In the U.S.), I'm in a good car, but I'm also proving that I can do a good job."

De Ferran, the pole sitter for today's race, turned down a lucrative offer to race in Formula One this year, deciding to return to Indy Car. In 1993-94, de Ferran became a star while racing in the Formula 3000 series, the minor-leagues of Formula One.

Last year, de Ferran won the season finale at Monterey, edging Christian Fittipaldi by just two points to become the Indy Car Rookie of the Year. In addition to Monterey, de Ferran also finished second at Vancouver, British Columbia, and had three other top-10 finishes.

"There's a lot of pressure involved in every part of Formula One," said de Ferran, who grew up in Sao Paulo, Brazil. "You've got to work extremely hard to get there and work even harder to stay there.

"It's better here because it's

Carlo because it's easy to gain confidence from people. The people are really nice here.

"I feel much more at home here than I ever did in England, and that's another reason I've enjoyed coming. Even if I could have returned to Formula One, I would have chosen this, anyway."

Last year, Zanardi raced with the Lotus Esprit team in the British Racing Production championship, winning the GT2 class at Silverstone and finishing fourth overall.

His career has been filled with ups and downs. After finishing sixth early in the Formula One season in 1993, Zanardi was sidelined by a major accident in Belgium that caused him to miss the rest of the season. After returning to Formula One as a test driver in 1994, his racing team folded.

Before racing Formula One, Zanardi dominated the Formula 3000 series in 1991. He qualified on the front row in 9-of-10

Photos by BOB CAREY / Los Angeles Times

Workman sprays white paint on concrete barrier along backstretch of Grand Prix racecourse in downtown Long Beach.

Wheels of Fortune

Long Beach Revs Up for Annual High-Profile, High-Profit Grand Prix as Workers Race to Transform City's Streets Into a Twisting Track

Michael Andretti

Gil de Ferran

Juan Fangio

Parker Johnstone

Robbie Gordon

Brian Herta

Adrian Fernandez

Greg Moore

1997

The 1997 Toyota Grand Prix of Long Beach, race three of the CART season, was held on Aprll 13, 1997 in front of a crowd of 90,000. Thursday was the Concours d'Promenade in downtown Long Beach featuring a children's grand prix in addition to race cars and classics on display.

Friday was practice and qualifying for all. Saturday featured the Toyota Celebrity Race at 1:45 PM and the Kool Toyota Atlantic Race at 3:45 PM.

Sunday, race day, had the Firestone Indy Lights Race at 10:30 AM with the CART Race at 1:00 PM followed by the North American Touring Car Race at 3:40 PM..

The race was live on XTRA Radio, the Kmart CART Radio Network and on ABC TV, tape delayed in the Los Angeles area. Danny Sullivan and Jack Arute were reporters and Bob Varsha was the announcer.

The Long Beach Grand Prix Association expanded its interest, buying Gateway International Raceway and Memphis Raceway, then sold shares to both International Speedway Corporation and Penske Motorsports. Dover Downs Entertainment then bought the remaining shares.

Nice day at the beach for de Ferran

By NATE RYAN
Sun Sports Writer

LONG BEACH — Forgive Gil de Ferran if he seemed a bit reserved after dominating the first day of qualifying at the Long Beach Grand Prix. He's been down this road before — with poor results.

De Ferran easily won the provisional pole Friday, turning a lap of 51.293 seconds with a top speed of 111.313 mph in his Reynard-Honda.

"It's good to be back here,

said. "It's only Friday."

Last year, de Ferran was the Long Beach pole-sitter after setting a track record during qualifying.

De Ferran's mastery of the track continued during the race, as he led the first 100 laps. But a blown exhaust hose knocked him out on Lap 101 with four laps remaining.

The memory of last year's debacle wasn't far from de Ferran's mind on Friday. The Paris native has experienced a string of bad luck through the first two

twice.

"What happened last year was very unfortunate," de Ferran said. "We've had a tough year and a tough week rebuilding the cars after the crashes. But this team has been fantastic through the tough patches."

De Ferran's return to form could spell trouble for Alex Zanardi's streak of six consecutive poles, a CART record. Zanardi was third during Friday's time trials with a lap of 51.843,

See LBGP/C6

■ Zanardi blasts CART, other drivers. Notebook/C6

Gil de Ferran had the pole but crashed and started in his backup car. Alex Zanardi won with Mauricio Gugelmin second and Scott Pruett third. De Ferran led the first 26 laps when a yellow prompted pit stops. Raul Boesel stayed out and led for 14 laps . De Ferran was leading when he crashed on lap 69. Vasser took over until Zanardi took the lead with 12 laps to go when Vasser pitted for fuel.

Alexander Tagliani won the Toyota Atlantic Race with Anthony Lazzaro second and Bill Auberlen third. Sean Patrick Flanery won the Toyota Celebrity Race with Jason Bateman second, Christy Carlson third and Tom Kendall, the first professional, fourth. Helio

Castroneves won the Firestone Indy Lights Race with Cristiano da Motta in second and Mark Hotchkis third. Neil Crompton won the North American Touring Car Race followed by Randy Pobst. and Dominic Dobson.

By Any Name, Zanardi a Rather Easy Winner

■ **Motor racing:** Italian victorious in uneventful Grand Prix of Long Beach by 3.8 seconds over Brazil's Gugelmin.

By SHAV GLICK
TIMES STAFF WRITER

Alex Zanardi, who would rather be known as Alessandro, had a feeling it was a good omen Friday when he went to pay his bill at a fancy Long Beach restau-

Alex Zanardi

Read the Autoweek article on the attached DVD and watch the 1997 video on our you tube channel - www.youtube.com/@racinghistoryproject2607

Flanery Drives Way to Celebrity Victory

By MIKE KUPPER
TIMES ASSISTANT SPORTS EDITOR

LONG BEACH—Actor Sean Patrick Flanery, to be seen soon in a movie called "Independence," showed plenty of it Saturday in winning the pro-celebrity race at the Toyota Grand Prix of Long Beach.

Starting on the pole, Flanery left the madding crowd far behind and drove to an uneventful victory in an otherwise eventful race, one memorable because of a three-car accident before the field got to the first turn, as well as a turf dispute in which Jack Brabham, former world Formula One champion, finished a battered second to local Trans-Am sedan series driver Tom Kendall.

It was all in fun—the drivers, in similarly prepared Toyota Celicas, were raising money for Southern California children's hospitals—but intense just the same.

"I was really aware of what was going on [behind me]," said Flanery, an accomplished amateur driver who plans to compete in a minor league pro series later this season. "Coming down the front cloud of smoke and a couple of cars emerging from it."

What they were emerging from was a melee touched off by actor Jason Bateman, a former celebrity winner at Long Beach, who suddenly turned left when he should have been going straight, taking out fellow actor Grant Show, the defending celebrity winner. Also caught up in the action was TV weatherman Dallas Raines.

Bateman went on to finish second and Raines continued on too, eventually finding himself in the midst of the Kendall-Brabham dispute. He raced with the veteran pros, who started 30 seconds behind the celebrities, for a lap or two, then wisely got out of the way. Good thing too, the way Kendall and Brabham were going at it.

Brabham held off the determined Kendall for several laps but when Brabham—Black Jack in his racing days, Sir Jack now—tried once more to shut the door on Kendall going into Turn 1, Kendall refused to yield and put Brabham's car into the wall.

"I don't think he's ever seen a

□

In the only pro race of the day, a demolition derby that made the celebrity set-to look like a model of racing decorum, Canadian Alexandre Tagliani turned back a late-race challenge by Anthony Lazzara of Acworth, Ga., and won the Kool-Toyota Atlantic event.

The race, for open-wheel cars powered by Toyota engines, was shortened from 38 scheduled laps to 30 after a pileup in the hairpin turn prompted about a half-hour interruption.

Canadian Bertrand Godin spun coming out of the hairpin, tried to right his car by going the wrong way on the track and was hit by another Canadian, Cam Binder, touching off the melee.

Trouble dogged the race, much of which was run under the caution flag.

□

Al Unser Jr., once the undisputed king of Long Beach with six Indy car victories since 1988, four of them in succession, apparently has fallen on evil days.

He qualified 15th in his Penske- completing only 27 of 147 laps because of electrical problems, then last weekend in Australia, he lost a wheel only 10 laps into the race.

"I'm just really frustrated right now," Unser said. "My crew and I have done everything to get the car working great for one quick lap so we could move up on the [starting] grid. Now I don't know what to do because we don't have a race setup. I've really painted myself into a corner."

Here last season, Unser qualified ninth, then finished third with a display of patient driving in another not-quite-right car. Winner of 31 CART races in his 15-plus-year career, he hasn't won since 1995 and has not sat on a pole since 1994.

□

Recent retiree, publicist Hank Ives of Orange, was honored at a track gathering for his 40 years in auto racing.

Ives, who served as publicity director of the Long Beach Grand Prix in its formative years as a

Grand Prix's Speed: From Fast to Faster

■ **Auto racing:** Five drivers beat course record during qualifying for Sunday's race in Long Beach.

By SHAV GLICK

Jones eyes 'hometown' win

Driver from Rolling Hills expects large rooting section

By Brian Patterson
STAFF WRITER

If the Toyota Grand Prix of Long Beach sets an attendance record this weekend, PJ Jones expects a call of thanks from race organizers.

"I've been getting calls for tickets every day," said Jones, 27, of Rolling Hills, who will try today to win what he calls his "hometown race."

"Chris Pook (president of the

"We've got him 85 to 90 percent back," PJ said, "and that includes his great sense of humor. We almost lost him, so considering that alternative, we're blessed."

PJ Jones

(today) is going to be something special. He's still real competitive, so maybe with him pushing me, we'll have a good showing."

PJ (whose given name is Parnell Velko) is in his first full year of driving for Dan Gurney's All-American Racers team, which is developing the Toyota engine for racing. The team has struggled this year, with Jones finishing 28th at Homestead, Fla., and 26th at Surfer's Paradise, Australia.

Auberlen vies for KOOL/Atlantic title

After a very successful start of the 1997 racing season in Florida, Bill Auberlen of Redondo Beach is planning to make his 1997 California debut equally successful at the Grand Prix of Long Beach. Auberlen, driving a BMW M3, took identical first in class and ninth overall in both the Rolex 24 hours of Daytona and the 12 Hours of Sebring sports car races plus a respectable seventh place finish in the 1997 inaugural Formula Atlantic race at Homestead, Grand Prix of Miami, Florida.

In Saturday's KOOL/Atlantic race, Auberlen will be driving a RALT 41, a highly competitive car compared to the older model 40 RALT he raced at Homestead. With newer equipment plus familiarity of the Long Beach course, a podium finish is quite possible. Formula Atlantic has been used as a spring board to both PPG CART World Series and Formula 1, including such notables as current and past PPG CART champions Jimmy Vasser, Jacques Villeneuve and Michael Andretti. The start of the Atlantic race might be the most exciting of all the races to be run this weekend. Unlike the PPG CART race and other support races that use a rolling start (moving start behind a pace car), the Atlantics use a standing start — as used in Formula 1.

The 38 lap KOOL/Atlantic race is scheduled to begin at 3:45p.m. on Saturday, April 12.

Bill Auberlen

Al Unser Jr.

Michael Andretti

Helio Castroneves

Mauricio Gugelmin

Adrian Fernandez

Dario Franchitti

Jimmy Vasser

Roberto Moreno

Zanardi and Ganassi

TOYOTA

1998

The 1998 Toyota Grand Prix of Long Beach, race three of the CART season, was held on Aprll 15, 1998 in front of a crowd of 87,500. Total attendance for the weekend was 213,000. Thursday was the Concours d'Promenade in downtown Long Beach featuring a children's grand prix in addition to race cars and classics on display.

Friday was practice and qualifying for all. Saturday featured the Toyota Celebrity Race at 2:15 PM and the Kool Toyota Atlantic Race at 4:00 PM.

Sunday, race day, had the Firestone Indy Lights Race at 10:30 AM with the CART Race at 1:00 PM followed by the Super Touring Race at 3:40 PM..

The race was live on ESPN with commentators Danny Sullivan, Jack Arute, Jin Beekhuis and Bob Varsha and on XTRA Radio. The big news, Dover Downs Entertainment purchased the Long Beach Grand Prix Association; no changes are anticipated.

It rained in Saturday's qualifying, the first time ever in the history of the race. Bryan Herta got the pole with Bobby Rahal second and Adrian Fernandez in third.

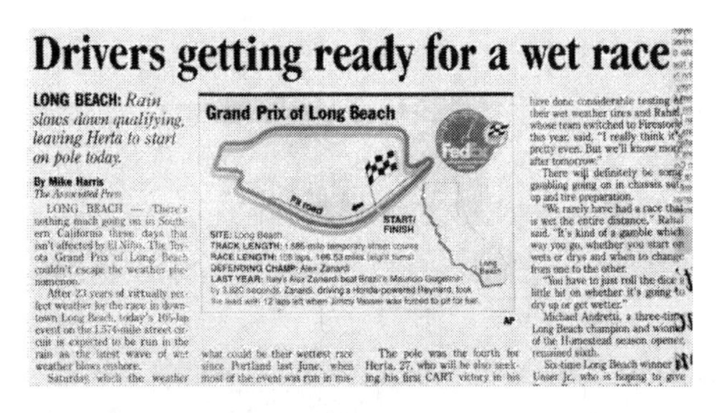

Drivers getting ready for a wet race

LONG BEACH: *Rain slows down qualifying, leaving Herta to start on pole today.*

By Mike Harris
The Associated Press

LONG BEACH — There's nothing much going on in Southern California these days that isn't affected by El Niño. The Toyota Grand Prix of Long Beach couldn't escape the weather phenomenon.

After 23 years of virtually perfect weather for the race in downtown Long Beach, today's 105-lap event on the 1.574-mile street circuit is expected to be run in the rain as the latest wave of wet weather blows onshore.

Saturday, which the weather ... what could be their wettest race since Portland last June, when most of the event was run in ...

The pole was the fourth for Herta, 27, who will be also seeking his first CART victory in his ...

... have done considerable testing ... their wet weather tires and Rahal, whose team switched to Firestone this year, said, "I really think it's pretty even. But we'll know more after tomorrow."

There will definitely be some gambling going on in chassis set-up and tire preparation.

"We rarely have had a race that is wet the entire distance," Rahal said. "It's kind of a gamble which way you go, whether you start on wets or drys and when to change from one to the other.

"You have to just roll the dice a little bit on whether it's going to dry up or get wetter."

Michael Andretti, a three-time Long Beach champion and winner of the Homestead season opener, remained sixth.

Six-time Long Beach winner ... Unser Jr., who is hoping to give ...

SLICK TRACK: Bryan Herta, of New Albany, Ohio, sends up a mist of water as he heads down the straightaway on Saturday durin final qualifying for the Grand Prix of Long Beach. Herta earned the pole position for today's race.

Herta, Rahal Lead Qualifying With Engine That Could

■ **Motor racing:** Six of top seven in provisional lineup at Long Beach use virtually unchanged power plant.

By SHAV GLICK
TIMES STAFF WRITER

There is an adage in motor racing: "New cars go fast, but old cars win races." The same might be said of engines this year in CART's FedEx championship car series.

Grand Prix of Long Beach at a Glance

■ **When:** Practice and qualifying today. Support races today and Sunday. CART race Sunday.

■ **The course:** A 1.59-mile, eight-turn circuit in downtown Long Beach.

... what I hired him for. On the other hand, as a driver, I want another shot at overtaking him.

"But, hey, if it rains Saturday, we're in good shape."

Gil de Ferran, who has been on the pole for the last two Long Beach races, was in the fastest Honda-powered car, qualifying fifth at 110.906.

"We were unfortunate that we had a little rain at the end of our session, and also a misfire, so we were able to do nothing with the second set of tires," he said. CART ...

On lap one Al Unser Jr and Scott Pruett came together. Pruett escaped unscathed but Unser was out. In fact, six cars had been sidelined by accidents by lap 37; Unser, Tracy, Christian Fittipaldi, Jourdain, Carpentier, Papis. Pruett and Zanardi got together forcing Zanardi to make a pit stop for repairs. Andrettii got airborne after hiiting Rahal but was able to continue for a short time, dropping out on lap 55. Gil de Ferran had a blown tire and pitted for a tire change on lap 12, regained the lead during a yellow on lap 40 and led until his lap 93 pit stop and gearbox failure put him out. Eleventh place qualifier Zanardi took the lead from Bryan Herta and went on to win with Franchitti in second and Herta in third.

Zanardi Steals the Show

■ **Motor racing:** In an incident-filled Grand Prix of Long Beach, Italian driver weaves his way back to take checkered flag.

By SHAV GLICK
TIMES STAFF WRITER

The rains failed to arrive, the sun came out and impetuous Alex Zanardi helped the 24th Toyota Grand Prix of Long Beach become the most exciting and competitive race since John Watson came from far back to win the final Formula One race in 1983.

CART FedEx champion Zanardi, after qualifying a disappointing 11th as defending Long Beach

GINA FERAZZI / L.A. Times

"I have no words to describe it," Alex Zanardi said of Grand Prix win.

■ **Commentary:** Zanardi again makes a super effort to overcome crashes, gridlock and bent equipment.

By MIKE KUPPER
TIMES ASSISTANT SPORTS EDITOR

How do you say Superman in Italian?

Zanardi, obviously. As in Alex Zanardi, winner of Sunday's Toyota Grand Prix of Long Beach.

Faster than a speeding bullet? You bet, if the speeding bullet is Bryan Herta.

More powerful than a locomotive? Sure, when the locomotive is a 10-car pileup in the tempo-

Cristiano da Matta won the Indy Lights Race followed by Geoff Boss and Naoki Hattori. Memo Gidley won the Toyota Atlantic Race with Andrew Bordin in second and Andrea de Lorenzei in third. Sean Patrick Flannery won the Toyota Celebrity Race and the professional class. Flannery was considered a pro from his previous win. The first amateur was Andy Lauer in second with John Cabe in third. The Trans Am Race was won by Paul Gentilozzi with Stu Hayner second and Johnny Miller third.

Long Beach-Based Grand Prix to Be Sold

By VANESSA HUA
TIMES STAFF WRITER

Racing into position, **Dover Downs Entertainment** Inc. on Friday agreed to buy **Grand Prix Assn. of Long Beach** Inc. in a $90-million stock-and-cash deal that would make Dover one of the biggest operators of motor sports in the U.S.

The combination would create an entertainment and sports powerhouse. It would also expand the Dover, Del.-based company to include ownership and operation of

DOVER: Grand Prix Sold for $90 Million

Continued from D1
demonstrating they can handle large crowds and other complex logistics, owners can attract prestigious and profitable events.

"You need a track record of success to get race dates from the sanctioning bodies," said Scott Barry, a leisure and entertainment analyst for Raymond James in St. Petersburg, Fla.

The company's flagship Delaware site currently draws fans from Philadelphia, New York and Washington. Dover also owns and operates Nashville Speedway, USA, in Nashville.

Dover would acquire the three courses and races owned and operated by Long Beach-based Grand Prix Assn., including the popular temporary street course that serves the annual Toyota Grand Prix of Long Beach. The 24th annual race will be held April 3 to 5 and is expected to draw more than 200,000 spectators over the three days.

Grand Prix is also the owner and operator of Gateway International Raceway in Madison, Ill., and Memphis Motorsports Park in Tennessee.

"We're pleased to be involved with a company with very little debt and good cash reserves," said Christopher Pook, Grand Prix chairman and chief executive.

Grand Prix zipped up 75 cents to close at $17.13 on Nasdaq on Friday, and shares of Dover dropped 31 cents to $28.75 on the New York Stock Exchange. Grand Prix went public at $10 a share in June 1996.

Under terms of the deal, Grand Prix shareholders would receive .63 of common stock in exchange for each Dover share, subject to adjustment if Dover's closing price is greater than $32 or lower than $21. Shareholders must approve the deal, which is expected to close in June.

As part of the agreement, Dover Downs also purchased **International Speedway** Corp. and **Penske Motorsports** Inc.'s combined interest of 14.6%, or 680,000 shares, in Grand Prix for $15.50 a share, or $10.5 million.

No Grand Prix employees are expected to be laid off once the merger is completed, the companies said. Grand Prix would retain its name and operations, and Pook would remain CEO.

Anti-tobacco groups target auto racing sponsorships

LONG BEACH (AP) — Anti-tobacco forces, having already marred the Marlboro Man and successfully lobbied to outlaw smoking in California bars, are targeting another bit of Americana: Cigarette sponsors of auto racing.

Three cars at this weekend's Long Beach Grand Prix are either sponsored or will be decaled with stickers from the American Cancer Society, an anti-tobacco group and the company that makes anti-smoking patches and gum.

"It's great," said anti-tobacco crusader Pat Etem, whose group L.A. Link is sponsoring a car in one of the weekend's support races for the Grand Prix. "I just can't wait to see it on the track."

For the first time, the American Cancer Society will have its logo on two Indy-type cars, and Etem's group is using Proposition 99 tobacco tax money to sponsor a Trans-Am. SmithKline Beecham, a pharmaceutical company, has recast its Indy-type car to play up its Nicorette gum and Nicoderm nicotine patches.

The cars will race against those heavily financed racing teams sponsored by the makers of Marlboro, Kool, Player's and Hollywood cigarettes.

Off the track, anti-tobacco activists could win big. A $368.5 billion agreement under consideration in Congress contains a provision barring them from sponsoring athletic events such as the Grand Prix.

Alan Henderson, a professor of health sciences at California State University, Long Beach and president of the state chapter of the American Cancer Society, said that even if cigarette companies leave auto racing, organizations like his are likely to stay.

"We think this is a great place for the American Cancer Society to be," said Henderson, a longtime racing buff.

Race car sponsors keep their budgets a closely held secret, but estimates are that the tobacco companies spend as much as $50 million collectively to sponsor cars in the Long Beach Grand Prix and other Indy-style races.

Anti-tobacco activists, like the cigarette companies, have discovered that sponsoring race cars is a great way to reach young people.

"Race cars are spontaneously attractive to kids," said

Alex Zanardi

Andre Rebeiro

Scott Pruett

Gil de Ferran

P.J. Jones

Hiro Matsushita

Greg Moore

Al Unser Jr.

THE ASSOCIATED PRESS

Paul Tracy, right, goes airborne as he and Christian Fittipaldi collide at the Long Beach Grand Prix.

CART FedEx Championship Series
Round 3 of 19
Toyota Grand Prix of Long Beach
Eight-turn, 1.59-mile
street circuit
Long Beach, Calif.
April 5, 1998

It must have been an interesting pre-race strategy meeting at Team Rahal. With Bryan Herta on the pole and car owner Bobby Rahal beside him, the afternoon looked promising for the winless, long-suffering Herta. And for the first 20 laps, Herta was coolly comforted to see his car owner in his mirrors.

Then, on lap 28, Rahal took himself out of the picture by bumping Michael Andretti in the hairpin and stalling. In the same encounter, Andretti tweaked his suspension; a tire ruined as a result let go on lap 55, and Andretti hit the wall while running a strong third. For his quicksilver driving to [...]

Alex Zanardi and his crew celebrate after his run from a lap down to top of the podium.

[...] overshot a turn.

Adrian Fernandez followed his win in Japan with a steady fourth, and left Long Beach with the points lead. ■

Results

1) Alex Zanardi, Reynard-Honda, 105 laps at [illegible] mph avg. speed; 2) Dario Franchitti, Reynard-Honda, 105; 3) Bryan Herta, Reynard-Ford, 105; 4) Adrian Fernandez, Reynard-Ford, 105; 5) Tony Kanaan, Reynard-Honda, 105; 6) Greg Moore, Reynard-Mercedes, 105; 7) Mark Blundell, Reynard-Mercedes, 105; 8) Jimmy Vasser, Reynard-Honda, 105; 9) Helio Castro-Neves, Reynard-Mercedes, 105; 10) Mauricio Gugelmin, Reynard-Mercedes, 105; [illegible]

Read the Autoweek article on the attached DVD and watch the video on our youtube channel - www.youtube.com/@racinghistoryproject2607

1999

The 1999 Toyota Grand Prix of Long Beach, race three of the CART season, was held on Aprll 18, 1999 in front of a crowd of close to 100,000. Some changes were once again made to the circuit, now 1.824 miles, with a left hand turn off Shoreline Drive and three additional corners, all due to the construction of the aquarium. The race, originally scheduled for 80 laps was stretched to 85 over fuel consumption concerns.

Dwight Tanaka, the vice president of operations, said he now had a budget of $1.8 million to provide 1800 concrete barriers, 21 grandstands with 63,000 seats, six pedestrian bridges, 12 television / photo towers and 61 VIP suites. Mario Andretti and Dan Gurney were co grand marshals.

The race was on ABC TV, tape delayed in the Los Angeles area. Commentators were Parker Johnstone and Jon Beekhuis with Paul Page as announcer.

This year's schedule included the usual, Friday practice and qualifying for all and Saturday's Toyota Celebrity Race at 2:15 PM and the Kool Toyota Atlantic Race at 4:00 PM. Sunday, race day, had the PPG Dayton Indy Lights Race at 10:20 AM with the CART Race at 1:00 PM followed by the Johnson Controls Trans Am Race at 3:45 PM.

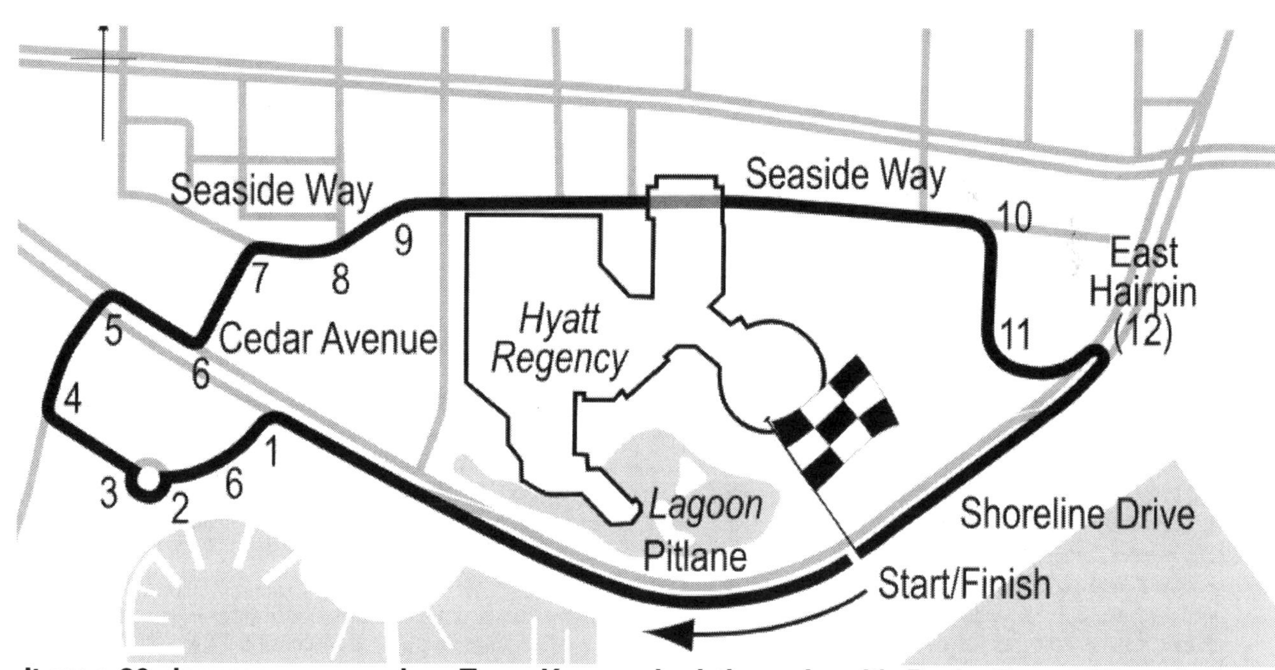

It was 89 degrees on raceday. Tony Kanaan had the pole with Dario Franchitti a close second but both started in their backup car. Bryan Herta took the lead, passing both Kanaan and Franchitti in turn one. Kanaan and Franchiti repassed Herta on lap two and built up a big lead. Montoya passed Franchitti on a restart after the Hattori caused yellow, pressured Kanaan until he made an error and crashed, leaving Montoya to win with Franchitti second and Herta third.

Kanaan takes pole at Long Beach GP

CLIFF KIRKPATRICK
STAFF WRITER

LONG BEACH — The Grand Prix of Long Beach turned 25 this weekend, and Tony Kanaan came of age Saturday.

Kanaan, a 24-year-old Brazilian, earned his first career pole position in this second season in a CART win and, eventually, a season championship.

Kanaan's breakthrough began with a new team and a major sponsor this year. He was hired by a different team from the cost effective Tasman Motorsports.

"Now that McDonalds has come aboard, I'm allowed to test," Kanaan said. "Now I've tested and have more money

(See all the race results on the attached DVD)

Montoya right on Target

LONG BEACH GRAND PRIX: *CART rookie wins series' first road race.*

By Mike Harris
The Associated Press

LONG BEACH — No doughnuts this time, just another checkered flag for Target-Chip Ganassi Racing.

After two straight victories by Alex Zanardi in the Toyota Grand Prix of Long Beach — and his gleeful celebrations that included spinning rubber doughnuts on the track — replacement Juan Montoya, a 23-year-old rookie from Colombia, blew away the field Sunday to earn his first CART FedEx Series victory.

Including a victory by teammate Jimmy Vasser in the 1996 race through the streets of downtown Long Beach, this was the fourth straight for Ganassi's three-time reigning series champions.

With Zanardi gone to Formula One, replacing a two-time series champion with a raw rookie seemed an unlikely formula for keeping the No. 4 Reynard-Honda at the front of the field. But confident Montoya, the 1998 European Formula 3000 champion,

Please see **GRAND PRIX** *on* **C9**

Still streakin'

A list of the last four winners of the Toyota Grand Prix at Long Beach, including team:

1996	Jimmy Vasser	Target-Ganassi
1997	Alex Zanardi	Target-Ganassi
1998	Alex Zanardi	Target-Ganassi
1999	Juan Montoya	Target-Ganassi

Alex Tagliani won the Toyota Atlantic Race with Anthony Lazzaro in second and Lee Bentham in third. Shaun Palmer won the Toyota Celebrity Race and was the first amateur. Roger Mears, the first professional was second followed by another amateur, Glen Plake. Phillip Peter won the Indy Lights Race followed by Scott Dixon and Geoff Boss. In the Trans Am Race, Paul Gentilozzi won followed by Johnny Miller and Michael Lewis.

Snowboarder Palmer Enjoys Smooth Ride

■ **Auto racing:** Xtreme Games gold medalist wins celebrity race.

By MIKE KUPPER

Schedule

■ **Today:** PPG-Dayton Indy Lights race, 10:20 a.m.; Toyota Grand Prix of Long Beach, 1 p.m.; Johnson Controls Trans Am race, 3:45 p.m.

Associated Press

Juan Montoya leads Dario Franchitti in straightaway during Grand Prix of Long Beach.

Hiroic Plans

"A winning Atlantic car is not enough. Developing an Indy car, and a successful Indy car, will be the best demonstration of our technology."

Read the Autoweek article on the attached dvd and watch the video on our you tube channel - www.youtube.com/@racinghistoryproject2607

Dario Franchitti

Jimmy Vasser

Michael Andretti

Greg Moore

Juan Pablo Montoya

Tarso Marques

Adrian Fernandez

Bryan Herta

ENTIRE STOCK
DOCKERS® KHAKIS
ON SALE*
DOCKERS KHAKIS
JOIN US
Robinsons-May and Dockers® are proud to be the presenting sponsors of "Fashion in the Fast Lane," a fashion show fundraiser for the Grand Prix Foundation. Join us April 15, from 11AM-2PM at the Westin Hotel Long Beach. For more information, call (562) 490-4509.
GRAND PRIX FOUNDATION
LONG BEACH
1999
CARA
GIFT WITH PURCHASE
Receive a Dockers® racing theme mouse pad with any $50 Dockers® Khakis purchase.
Try on any pair of Dockers® Khakis and receive driver Helio Castro-Neves' Hero Card.
SALE
39.99
ALL DOCKERS® 5-POCKET KHAKIS
Orig. $50
In khaki, stone, olive and black.
SALE
39.99
ALL DOCKERS® AUTHENTICS KHAKIS
Orig. $50
HOGAN
Firestone
*Excludes Dockers® K1 khakis. Original prices are offering prices only and may or may not have resulted in sales. Advertised merchandise may be on sale in the future. Dockers® is a registered trademark of Levi Strauss & Co.
CHARGE IT WITH ROBINSONS-MAY
WE ALSO WELCOME VISA
MASTERCARD AND DISCOVER
ROBINSONS · MAY
THIS IS WHERE IT GETS GOOD
ORDER ANYTIME
CALL TOLL-FREE
1-800-633-1224

GRAND PRIX - GRAND SHIP
ANOTHER WINNER - THE QUEEN MARY
Long before Indy cars were speeding through the streets of Long Beach, the Queen Mary was racing from Southampton, England, to New York City in record time - and providing luxury service all the while. In fact, for 14 years, from 1938 to 1952, she held the Blue Riband, an honor awarded to the fastest ship at sea! So while you're here, ride the free Passport Shuttle from downtown Long Beach to the Queen Mary, or hop aboard the AquaBus water taxi, and come aboard. Tour the ship on your own, or book a guided tour, enjoy an elegant-to-casual meal, tour the Russian submarine, Scorpion; visit our many shops for souvenirs, or even book a stateroom, and plan to spend the night. She may no longer sail, but the Queen Mary will always be a winner!
SIR WINSTON'S - Open daily 5:30 p.m. to 10:00 p.m. Gold Award-winning California and Continental cuisine including succulent cuts of beef, game, fresh seafood, poultry, pasta specialties and exotic desserts. Panoramic harbor views. Reservations required.
THE CHELSEA - Lunch served Monday thru Friday, 11:30 a.m. to 2:00 p.m. Dinner Wednesday thru Sunday, 5:30 p.m. to 10:00 p.m. Gold Award-winning fresh seafood specialties, such as Seafood Phyllo and Cioppino are the Chelsea's hallmark. Reservations required.
CHAMPAGNE SUNDAY BRUNCH - Served from 10:00 a.m. to 2:00 p.m. in the Grand Salon. Race day and every Sunday, feast on this array of international favorites. Food stations plus live music and a children's buffet. Reservations recommended.
SPECIAL GRAND PRIX QUEEN MARY TOUR SAVINGS
BRING THIS AD AND SAVE $3.00
Present this ad at the Queen Mary Box Office and save $3.00 per person, all ages, for up to 4 people. May not be combined with any other offer. Regular General Admission prices, $15.00 for adults, $13.00 for seniors and military and $9.00 for children ages 4-11. The Scorpion Submarine Tour, $10.00 for adults and $9.00 for children, is not included in General Admission. PLU A16 C17 S18
FOR RESERVATIONS, CALL (562) 435-3511
World Wide Web: www.queenmary.com E-Mail: queenmry@gte.net

2000

The 2000 Toyota Grand Prix of Long Beach, race three of the CART season, was held on Aprll 16, 2000 in front of a crowd of over 100,000. Thursday was the Concours d'Promenade in downtown Long Beach featuring a children's grand prix in addition to race cars and classics on display.

Friday was practice and qualifying for all. Saturday featured the Toyota Celebrity Race at 1:45 PM and the Kool Toyota Atlantic Race at 3:45 PM

Sunday, race day, had the Firestone Indy Lights Race at 10:30 AM with the CART Race at 1:00 PM followed by the Super Touring Race at 3:40 PM.

The race was live on XTRA Radio, the Kmart CART Radio Network and on ESPN with commentators Paul Page, Parker Johnstone and Jon Beekhuis. Getting in the way all weekend, Sylvester Stallone's film "Driven" was filming and using scenes from the race.

de Ferran wins pole, but seeks a greater payback

▶ **AUTO RACING:** He's lost out on several good chances to win Grand Prix of Long Beach.

By MIKE HARRIS
Associated Press Writer

LONG BEACH — If ever a race owed a driver a victory, it's

MORE AUTO RACING
■ **BUSCH SERIES:** Nemechek's victory is pretty as a picture. / Page C8

105 laps before being victimized by a disconnected turbocharger hose that ended his day and gave the win to Vasser.

Asked what kind of a race he

TOP QUALIFIERS
LONG BEACH (AP) — Top qualifiers for today's Toyota Grand Prix of Long Beach, with car number in parentheses, driver, type of car, time in seconds, speed in mph:
1. (2) Gil de Ferran, Honda Reynard, 67.494, [illegible].
2. (12) Jimmy Vasser, Toyota Lola, 67.703, 104.645.
3. (1) Juan Montoya, Toyota Lola, 67.984, 104.213.
4. (40) Adrian Fernandez, Ford Reynard, 68.001, 104.178.

er Racing to Marlboro Team Penske — also will change his Long Beach luck.

Although the Penske team has fallen on hard times the past few seasons, Roger Penske's drivers have won this race four times — Danny Sullivan (1992), Paul Tracy (1993) and Al Unser Jr. (1994-95).

A victory by de Ferran would end Target/Chip Ganassi Rac-

Gil de Ferran had the pole but Paul Tracy won, having started 17th, with Helio Castroneves in second and Jimmy Vasser third. Only nine drivers finished on the lead lap. De Ferran led by three seconds when he pitted on lap 30 and exited the pits in 18th pace. He then had a collision with Vasser, damaged his front wing and was not in contention after that.Tracy and Andretti collided in the pit lane with no apparent damage while Christian Fittipaldi had a pit fire that ended his race.

Perfect set-up brings Tracy a surprise Long Beach GP triumph

LONG BEACH (AP) — When Paul Tracy went to bed Saturday night, he wasn't feeling very confident.

Not only had he qualified 17th for the second time in the young season, his Reynard-Honda wasn't handling well enough to give the 31-year-old Canadian much hope for the Toyota Grand Prix of Long Beach.

That all changed Sunday, thanks to new Team Kool Green engineer Steve Challis.

Challis previously worked with Greg Moore, killed last year in the season-ending race in Fontana.

"We were just lost all weekend," Tracy explained after driving to the 16th victory of his CART FedEx Series career and second in the streets of Long Beach. "We'd exhausted all the setup ideas we had, all our notes from previous races and we weren't really getting anywhere.

"Then Steve broke out one of Greg's notebooks from last year and we put on one of their setups and the car was just perfect."

His performance was the farthest back a winner has started on a temporary street circuit since Al Unser Jr. won from 19th in 1986 on the old downtown course in Miami.

"That's two races now we've started in the back, but there's no panic on this team," Tracy said. "Starting 17th, the plan for us was just to stay alive and not get in a wreck.

"Being a lot less patient in the past, I would put my nose in where it shouldn't be and get it chopped off. This time, I wasn't being overaggressive and we stuck to our plan."

It worked to perfection for Tracy, who avoided most of the trouble that knocked out all but 12 of the 25 starters in the 82-lap event race. The win, combined with a third-place finish in the season-opener in Homestead, Fla., moved Tracy, who never has finished better than third in the series, into the points lead.

Tracy's only close call came in the pits on lap 11 when several of the drivers made their first stops during the first of six caution periods on the 1.968-mile, 11-turn circuit winding around the downtown convention center and along Long Beach harbor.

Michael Andretti darted out of his pit just ahead of Tracy, who bumped the rear of Andretti's car. But neither sustained any serious damage.

With teams trying a variety of fuel strategies on the tight course, team owner Barry Green's crew chose to go with three pit stops and watch the race unfold. It worked as Tracy took over lead on lap 62, passing rookie Takuya Kurosawa, the first Japanese driver to lead a CART event.

Helio Castro-Neves and 1996 Long Beach winner Jimmy Vasser had an intense battle for second during the final 18 laps, with Castro-Neves holding him off by half a car length for second.

Vasser, who finished fourth in Homestead, is second to Tracy, trailing in the points 34-26 after two of 20 races.

Foreign Invasion Considered a Big Plus on Circuit

MOTOR RACING

SHAV GLICK

When travel agent **Chris Pook** first broached the idea of a Formula One race through the streets of Long Beach 26 years ago, his hope was to spread the word that the sleepy beach town was worthy of international recognition. The Queen Mary had recently dropped anchor there and Pook wanted to make the Long Beach Grand Prix something of a Monte Carlo West.

After a few years, financial obligations caused him to drop F1 and run the more cost-conscious Indy cars with a nearly all American cast.

This year, in the 26th renewal, Long Beach will experience *deja vu*.

Of the 25 drivers in Sunday's CART FedEx race, only four are Americans. And two of the four are only substitutes for injured foreign drivers.

Michael Andretti, at 37 the senior member of the U.S. group, says the trend toward a more international look in CART is a plus, not a negative, as some perceive.

"Times have changed," Andretti said. "Most sports have become more international, so CART and champ cars are only following the trend. Look at hockey, baseball and the Masters last week. There are foreigners

Andretti.

It was on the twisting street course that he scored his first Indy car win, in 1986. He has won 37 more races since then—most by any active driver—but no more at Long Beach, despite having started on the pole in 1991, 1992 and 1995.

"It's always nice to go back where you've won, but more than that, Long Beach is a special place," he said. "Chris Pook does a great job. It's the way all events should be run. It's cool.

"I think we've helped Long Beach too. The area around the circuit has changed unbelievably since we started racing here. The buildings on the front straight used to be boarded up X-rated theaters and today it's a nice metropolitan area. I think the race had a lot to do with the rebirth of the downtown area."

For four years, from 1984 to 1987, the Long Beach Grand Prix was an Andretti happening. Michael's father, **Mario**, won in 1984 and 1985, Michael won in 1986 and Mario came back in 1987. Mario also won in 1977 when it was still a Formula One race.

The younger Andretti says he is hungry for another family win and that this might be the year.

"I'm especially looking forward to this week because we've really turned our program around," he said. "We're a lot better off than we were last year. We have gone back to Lola

"I went [to Indy] with nothing, hoping to find something," he said. "I spent day after day sitting on the wall, walking through the garages, hoping someone would see me. I had been there nearly a week and the rookie tests were about over when I got a call Sunday night about 10:30 that I could test the next morning at 11."

Without having driven on an oval since a Toyota Atlantic race in 1998 in Milwaukee, Gidley completed the entire 40-lap test, plus 21 practice laps, in little more than an hour. He drove for Team Pelfrey in the same car he will try to qualify next month for the 500.

"Indy was awesome," he said. "I thought I was going fast when I warmed up at 135 mph, but when I got to 210 I knew I was going fast."

Gidley drove in 10 CART races last year, six with **Dale Coyne** and four with **Derrick Walker**, but none were on ovals.

It is ironic how he landed with Forsythe's team. When Japanese rookie **Shinji Nakano** was injured in testing one of Walker's cars, Forsythe lent **Bryan Herta**, his third driver, to Walker for Sunday's race. So when Carpentier, Forsythe's No. 1 driver, was injured in a fall Monday, Herta was unavailable and the call went out to Gidley.

EIGHT FOR EIGHT

There have been eight Winston Cup races

The Associated Press

A LITTLE CROWDED: Long Beach Grand Prix winner Paul Tracy moves through a series of tight turns.

Scott Dixon won the Indy Lights Race ahead of Jason Bright and Felipe Giaffone. Tomy Drisi won the Trans Am Race followed by Willy T. Ribbs and Bob Ruman . The Toyota Celebrity Race was won by amateur Josh Brolin with Joshua Morrow second and Don Simons third. The first professional was Derek Daly in fourth. Buddy Rice won the Toyota Atlantic Race with David Rutledge second and Andrew Bordin in third.

American Monaco?

Read the Autoweek article on the attached DVD and watch the video on our youtube channel - www.youtube.com@racinghistoryproject2607

Josh Brolin follows father's career track

Josh Brolin is following in his father's treadmarks.

The actor won the 24th annual Toyota Pro-Celebrity Race on the streets of downtown Long Beach on Saturday — 22 years after his father, actor James Brolin, won the same race.

The younger Brolin outpaced a field of 16 other celebrity and professional drivers, including **Ashley Judd**, **George Lucas**, **Melissa Joan Hart** and **John Elway**.

The 10-lap celebrity race was held on the second day of the Toyota Grand Prix of Long Beach. The three-day event will be televised later this month on ESPN.

TEXACO

TOYOTA

AUTO CLUB

AUTO CLUB

TARGET
AUTO CLUB

Associated Press

Juan Montoya leads Dario Franchitti in straightaway during Grand Prix of Long Beach.

2001

The 2001 Toyota Grand Prix of Long Beach, race two of the CART season, was held on Aprll 8, 2001 in front of a crowd of over 90,000. Friday was practice and qualifying for all and a concert by "Chris Gaffney and Cold Hard Facts".

Saturday featured the Toyota Celebrity Race at noon and the Kool Toyota Atlantic Race at 4:30 PM and another concert, this featuring "Third Eye Blind".

Sunday, race day, had the Texaco Dayton Indy Lights Race at 11:25 AM, a noon concert by "One World", the CART Race at 1:00 PM followed by the Trans Am Race at 3:45 PM..

The race was on ABC with commentators Scott Pruett, Parker Johnstone and Jon Beekhuis. Paul Page was the announcer.,

There was on and off rain on Friday and Saturday but Helio Castroneves dodged raindrops and grabbed the pole as the track dried with Kenny Brack second fastest.

Castroneves shines in shootout for Long Beach pole

By MARK A. PEINADO
Staff Writer

LONG BEACH — If the last five minutes of qualifying are any indication, then the 27th annual Toyota Grand Prix of Long Beach could turn out to be the best ever run.

The pole position changed hands five times among three drivers and was being swapped by competing cars just as fast as the weather was changing from cold, gray morning skies to a bright, sunny afternoon.

Associated Press
Helio Castroneves heads out on rain tires during qualifying for the CART Toyota Grand Prix of Long Beach. Castroneves won the pole at 103.343 mph.

Helio Castroneves of Marlboro Team Penske finally emerged the victor as he clocked in at 103.343 mph. He rounded the 11-turn, 1.968-mile temporary street course in 1 minute, 8.55 seconds. Kenny Brack will start second and Tony Kanaan emerged in third.

"I'm very happy to have won the pole today, considering we had a rough start to the weekend," Castroneves said. "Yesterday, we went up and down trying to find the set-up and this morning we had to start all over because of the weather."

The weather proved to be a major factor in qualifying. When the first group of cars went out to turn their fastest laps, it was cold and the skies looked like they were going to burst. Mother Nature did let loose with some rain in certain parts of the track causing some teams to go to rain tires while others stayed on dry-weather slicks.

Michel Jourdain Jr. had the early pole under those conditions. It appeared Jourdain, in the Herdez Bettenhausen car, would get his first pole of his career. But as the second group of qualifiers went out, the sun slowly began to peek out. No one in

Please see LONG BEACH, C3

Pole sitter's career is skyrocketing

By MARK A. PEINADO
Staff Writer

LONG BEACH — He's known to most CART fans as the fence jumper because each time he has won a race, he's sprinted to the frontstretch grandstands and tried to climb the fence in celebration of a victory.

Helio Castroneves has accomplished that three times, and in the 2001 FedEx championship season, he's looking to become the next Spiderman.

"I hope to be No. 1," Castroneves said.

He's well on his way as he captured his fifth pole of his career on Saturday at the 27th annual Toyota Grand Prix of Long Beach. Castroneves edged Kenny Brack as time expired in the qualifying session.

Castroneves, 25, is expected to be the main challenger to Gil de Ferran in the fight to be this

Castroneves

year's CART champion. It is not easy challenging a teammate, but Castroneves is comfortable battling his friend.

"I have a lot of fun with him," Castroneves said. "I learned a lot from Gil."

As fast as Castroneves rounds a two-mile superspeedway, that's how quickly his career has skyrocketed in CART.

It was only four years ago, Castroneves was driving an Indy Lights car. He moved to CART in 1998 with Bettenhausen Motorsports. He finished second in the rookie of the year standings to Tony Kanaan. In 1999, Castroneves was with Hogan Racing. In that year, he captured his first pole and finished in the top-10

Please see CASTRONEVES, C3

Castroneves led every lap, ahead of Cristiano da Matta and Gil de Ferran. Michael Andetti had his motor quit; Zanardi had mechanical problems and Brack's gearbox broke after running second.. There were five caution periods; Garcia Jr, crashed on lap 10, Takagi and Gugelmin collided on lap 30, Carpentier and Papis collided on lap 55 and Takagi crashed again on lap 75. Castroneves was able to pull away at each restart and maintained his lead each time, collecting $100,000 for the win.. As has become his trademark, he climbed the fence in front of the fans before reporting to the winner's circle.

BRIAN WALSKI / Los Angeles Times

ong Beach Grand Prix winner Helio Castroneves sprays champagne at teammates after being hoisted by a member of his pit crew soon after finish.

CART at Long Beach/IRL at Homestead

Peace in Our Time... Perhaps

Read the Autoweek article on the attached DVD and watch the video on our youtube channel - www.youtube.com@racinghistoryproject2607

Nichola Minassian

Kenny Brack

Helio Castroneves

Tony Kanaan

Scott Dixon

Paul Tracy

David Rutledge won the Toyota Atlantic Race with Hoover Martins second and Joey Hand third. Scott Pruett won the Toyota Celebrity Race and was also the first professional, followed by Josh Brolin and amateur and Toyota dealer Tom Rudnai. Townsend Bell won the Indy Lights Race followed by Dan Wheldon and Derek Higgins. Lou Gigliotti won Trans Am Race with Johnny Miller second and Michael Lewis third.

Adrian Fernandez's rise to top wasn't easy: "Everything I have, I earned, because I had absolutely nothing."

HIGH-SPEED CHASE

Brazilians Leave Drama Out of Dance

MOTOR RACING
MIKE KUPPER

Vasser Proves He Still Has Some Run Left

MIKE KUPPER

Jimmy Vasser, driving a Toyota-powered car after switching from Honda, finished third at Long Beach Grand Prix.

Newman-Haas Goes International

■ **Grand Prix:** Team hands the keys to Fittipaldi and da Matta, two Brazilians who are poised to make an impact at Long Beach event.

By SHAV GLICK
TIMES STAFF WRITER

For the first time in 19 years, there will be no Andretti in the driver's seat of a Newman-Haas car in Sunday's Toyota Grand Prix of Long Beach.

Mario, a four-time winner, has retired, and Michael, a one-time winner, was dropped last year before hooking on with Barry Green's team.

In their place are two Brazilian pals, holdover Christian Fittipaldi and newcomer Cristiano da Matta.

Da Matta, in his first outing as Andretti's replacement, won the CART opener last month on a road course in Monterrey, Mexico. Driving a Toyota-powered Lola, he defeated defending series champion Gil de Ferran after battling former Indy 500 winner Kenny Brack most of the race.

That gave Newman-Haas a two-race winning streak, since Fittipaldi had won the final 2000 race at California Speedway.

Cristiano da Matta of Brazil will seek second consecutive CART victory in Sunday's Long Beach Grand Prix.

Race Facts

TODAY

- **Noon**—Toyota Pro/Celebrity Race
- **1**—Dayton Indy Lights Championship Qualifying
- **1:45**—Toyota Long Beach Grand Prix Qualifying
- **3:15**—Toyota Atlantic Championship Race
- **5**—Trans-Am Series Qualifying

SUNDAY

- **10:15 a.m.**—Texaco Havoline Challenge Dayton Indy Lights Championship Series Race
- **12:15 p.m.**—Toyota Long Beach Grand Prix
- **2:45**—Trans-Am Series Race

Bold Move by Rutledge Results in Narrow Win

By MIKE KUPPER
TIMES ASSISTANT SPORTS EDITOR

Canadian David Rutledge and Brazilian Hoover Orsi provided a tasty appetizer Saturday to today's Toyota Grand Prix of Long Beach.

Pole sitter Rutledge won a 32-lap race over the 1.968-mile street course for open wheel Toyota Atlantic cars but only after regaining a lost first-lap lead, then holding off a relentless Orsi until a gutsy move late in the race gave him a three-second cushion.

Unable to shake Orsi on his own, Rutledge passed the Santa Barbara Mazzotta brothers, Hawk and Zac, on the 28th lap, putting their lapped cars between his and Orsi's, surviving a wheel bump by Hawk Mazzotta in the process.

"I don't know if he saw me on the inside like that," Rutledge said of the tap.

Early on, Rutledge had to make up lost ground. Timing the starter's green flag perfectly in the season opener, Rutledge got a nice jump but then was passed

Instead, both she and Rudnai went into broadslides coming out of Turn 8, making things easy for the charging Pruett. He shot past them and so did actor Josh Brolin, last year's winner, who moved out of the celebrity class and into the pros this time around.

Pruett's winning speed was 63.452 mph, slower than he is used to running here but he was content.

"I'm amazed how competitive this race is," he said. "Josh is a fantastic talent. . . . At the end, it just came down to split-second decisions and I was able to take it. This is just a whole lot of fun."

Brolin finished second, and Rudnai got straightened out before Torres, beating her out for third.

□

With rain falling intermittently during the second round of qualifying, PacWest driver Mario Dominguez of Mexico sat on Friday's top speed and will start on the pole today in the Dayton Indy Lights 38-lap race.

Dominguez qualified Friday at 93.260 mph and made no attempt to better it, going out late in the Sat-

Team Rahal Recruits Talent From County

■ **Motor racing:** Mark Johnson, a former member of Rancho Santa Margarita-based PPI, brings along several top engine specialists.

By MARTIN HENDERSON
TIMES STAFF WRITER

Bobby Rahal, right, has hired Mark Johnson as director of operations to help driver Max Papis, left.

Castroneves, da Matta and de Ferran

2002

The 2002 Toyota Grand Prix of Long Beach, race two of the CART season, was held on Aprll 14, 2002 in front of a crowd of over 100,000. The Automobile Club Lifestyle Expo featured the "Gravity Tour", a collection of skateboarders, BMX bikes and inline skaters plus

Friday was practice and qualifying for all while Saturday featured the Toyota Celebrity Race at noon, the Chevron Challenge Toyota Atlantic Race at 4:30 PM and the Pioneer Rock 'n Roar concert with the "Goo Goo Dolls."

Sunday, race day, had the the popular driver introduction laps in Toyota Tundras, the CART Race at 12.30 PM followed by the Johnson Controls Trans Am Race at 3:45 PM.

The Grand Marshal was Long Beach Grand Prix founder and former president Chris Pook, who resigned to take the job as president of CART..

Qualifying was on the Speed Channel with race coverage on Fox with reporters Scott Pruett, Tom Kendall, Derek Daly, Calvin Fish and Jeremy Shaw and announcers Bob Varsha and Leigh Diffey.

Practice, qualifying and the support races for Sunday's 28th Toyota Grand Prix of Long Beach will be run over this 1.97-mile, 11-turn temporary street course around the Long Beach Convention and Entertainment Center and the Aquarium of the Pacific. The start-finish line is on Shoreline Drive, where the cars reach their highest speed.

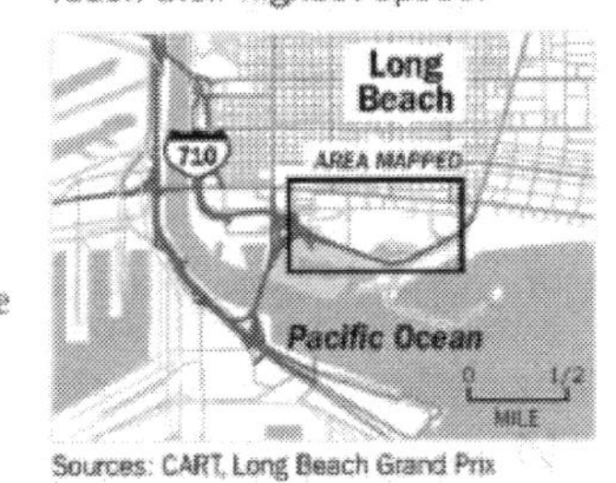

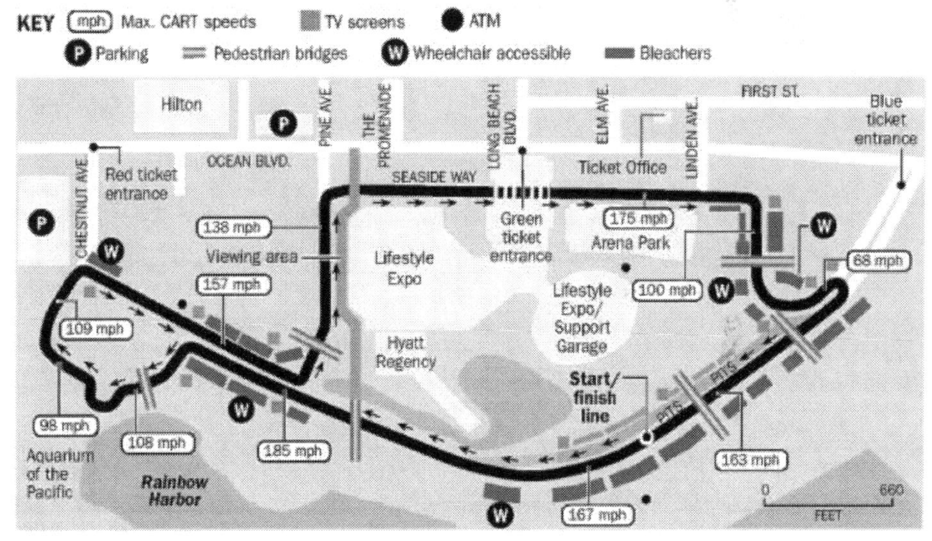

Sources: CART, Long Beach Grand Prix

Jimmy Vasser had the pole but finished second. Michael Andretti won, collecting $103,000. Jimmy Vasser second and Max Papis in third. Cristiano da Motta led for 17 laps, Kenny Brack grabbed the lead for one lap on lap 18 but de Matta regained it immediately. Kenny Brack led again for one lap until Michael Andretti took over from laps 33 to 46. Jimmy Vasser took the lead on lap 47, Brack had one more leading lap and them Andretti took over for good on lap 62. There were six cautions caused by Tony Kanaan on lap 17, Scott Dixon on lap 31, Bruno Junqueira on lap 52, Tagliani and Bell on lap 57, Dominguez on lap 63 and Servia on lap 81.

Andretti, Vasser lead 1-2 American sweep

ASSOCIATED PRESS

LONG BEACH — The only person not surprised when Michael Andretti found himself out front after his final pit stop in the Long

"Michael and Max had a funky strategy. It just shows you the fastest car doesn't always win."

Andretti, who had to go to a backup car after crashing in Saturday's qualifying, had Vasser

The Toyota Celebrity Race was won for the first time by a woman, Dara Torres, who was also the first amateur. Second was Christopher Masterson with the first professional, Danica Patrick in third. The Toyota Atlantic Race was won by Michael Valiante with Alex Gurney second and Jon Fogarty third. Paul Gentilozzi won the Trans Am Race followed by Boris Said and Johnny Miller.

(All The Race Results Are On the Attached DVD)

Swimmer Torres Makes a Splash

By MARTIN HENDERSON
TIMES STAFF WRITER

Saturday was Ladies Day at the Toyota Grand Prix of Long Beach.

For the first time in the 26-year history of the Toyota Pro/Celebrity race, a woman took the checkered flag.

Swimmer **Dara Torres**, an Olympic gold medalist, started from the pole and ran away from the rest of the celebrity field, doing something that previous stars **Cameron Diaz**, **Ashley Judd** and **Crystal Bernard** could not.

Torres wasn't alone. **Danica Patrick**, a 20-year-old from Illinois who raced in Europe the last few years, won the professional category and finished third overall despite starting 30 seconds behind the leader on the 1.97-mile street course.

Racing in identically prepared Toyota Celicas, Torres pulled away from No. 2 qualifier **Bill Goldberg**, a former wrestler who completed only four laps before crashing, and led by more than six seconds going into the last of 10 laps. Pumping her arm out the window before taking the checkered flag, Torres won by 3.31 seconds over actor **Christopher Kennedy Masterson**.

"She's a really solid driver who goes fast, brakes late and makes very few mistakes," said Masterson, the oldest

12 celebrities and, despite the starting handicap, finished 8.06 seconds behind Torres. Four-time Trans-Am champion **Tommy Kendall** was fourth, and two-time Indy 500 starter **Sarah Fisher** was fifth, giving women three of the top five positions.

"I don't think it has anything to do with women," Patrick said. "We were all drivers out there doing what we were brought here to do.

"The car doesn't know the difference between a girl or a guy."

Kendall, the most accomplished of the five professional drivers, said that celebrities are getting such good instruction from **Danny McKeever's** Fast Lane Driving School, and the race is so short, the professionals are going to have a tough time winning again.

"Without yellow flags, it's going to be rarer and rarer for the pros to win overall," he said.

Ford and Honda might be out of CART in 2003, but MG Rover announced Saturday that it is in. The English manufacturer is the first to confirm participation under CART's new rules calling for a 3.5-liter normally aspirated V8 engine. Cosworth, the current manufacturer of Ford's racing engines, has indicated it will continue in the series.

tive developments I have seen since I joined the organization [in December]," said CART President and Chief Executive **Chris Pook**. "As a native Englishman, I am honored to have one of Great Britain's premier brands join CART."

CART began racing last season in England.

Dorricott Racing won two of the last three titles in the defunct Dayton Indy Lights Championship—a step above Toyota Atlantics in CART's ladder system—and opened this season with a victory in the rain at Monterrey, Mexico, with rookie driver **Jon Fogarty**.

But veteran Atlantic teams threw down the gauntlet Saturday during qualifying, relegating **Bob Dorricott's** racing bloc to fourth, fifth and sixth on the starting grid.

Drivers for veteran Atlantic teams held the top positions as **Joey Hand** qualified first, **Michael Valiante** second and rookie **Ryan Hunter-Reay** third.

Hand, driving for DSTP Motorsports, averaged 90.494 mph to win the pole, and Valiante of Lynx racing averaged 90.463.

However, Hunter-Reay of Hylton Motorsports guaranteed himself a spot on the front by being the top qualifier during Friday's session.

ship up and know it's for real."

Fogarty will start fourth, **Alex Gurney** fifth and rookie **Luis Diaz** sixth for Dorricott.

Paul Gentilozzi and **Boris Said** started on the front row last year at Long Beach, and will do the same today in the Trans-Am series' Johnson Controls 100. Last year, the two made contact, and Gentilozzi's day ended on the first lap.

Gentilozzi, a three-time Long Beach winner, will start from the pole after averaging 86.854 mph, ahead of Said's 86.594. Both drivers are well aware of the perils of racing on the street.

"This is a race of attrition," said Gentilozzi, a three-time series champion. "It's easy to make a mistake here, clip a wall, flat-spot the tires. It's easy to get caught up emotionally."

That's what happened to Said last year.

"I could have won if I didn't let my ego get in the way," Said said. "I watched Paul put it in the fence, and then I was driving as fast as I could every lap for no reason. I could have backed off 10 or 15%."

And then Said put his car into the wall.

Lou Gigliotti went on to win. He will start fifth, behind **Butch Leitzinger**,

CART

The Race Report

CART

Read the Autoweek article on the attached DVD and watch the video on our you ube channel - www.youtube.com@racinghistoryproject2607

Adrian Fernandez

Alex Tagliani

Jimmy Vasser

Patrick Carpentier

Townsend Bell

Shinji Nakano

Cristiano da Matta

Michel Jourdain

Andretti, Vasser and Papis

Tent Companies Tap Into a Suite Kind of Hospitality

Events: More firms are offering executives and guests a privileged experience in outdoor corporate settings.

By ELENA GAONA
TIMES STAFF WRITER

Amid furiously racing cars, greasy tires getting changed and dust flying around the pit stops of the Long Beach Grand Prix, the event's outdoor hospitality suites were a plush, elegant oasis

He Prefers Being Graded on Curves

Motorsports: Da Matta, gunning for his fourth consecutive victory, likes the Long Beach course.

By SHAV GLICK
TIMES STAFF WRITER

Cristiano da Matta has won the last three CART races—the last two in 2001, including Fontana, and the season opener in Monterrey, Mexico—but don't ask him if he has been dominating.

"I've just won a few races," the jockey-sized Brazilian said after arriving in Long Beach for Sunday's 28th annual Toyota Grand Prix. "Dominating is the way Alex Zanardi or Juan Montoya won their championships. If I win a few more, maybe you can say that about me, but not yet."

Zanardi won five races in 1997 and six in 1998. Montoya won seven in 1999.

Da Matta, who drives a Chevron-sponsored No. 6 Toyota-Lola, has proved to be one of CART's most

Long Beach Grand Prix

When: Friday-Sunday.
Grand Prix: Sunday, 90 laps, 12:30 p.m.
TV: Channel 11.
Defending champion: Helio Castroneves.
Current points leader: Cristiano da Matta.
Qualifying: Friday, 1:40 p.m.; Saturday 1:45 p.m.
Support races: Saturday: pro celebrity, noon; Sunday: Toyota Atlantic, 10 a.m.; Trans-Am, 3:45 p.m.

straight stretches and some difficult turns, so we must struggle to find a compromise."

Da Matta finished second last year, splitting Penske teammates Helio Castroneves and Gil de Ferran. Castroneves led all 82 laps, but Da Matta was right on his bumper

dent. One eliminates fuel economy races and the other makes qualifying more equal and profitable.

"Every driver, every team, even the engineers, used to complain about having to conserve fuel during a race. It was like they would say, 'We're spending all this money, all these engineers working here, to make all this power, then it comes race time, sometimes we have to lean the engine to get better fuel economy for race strategy.'

"So the fact that we are going to be able to run as fast as you can, using as much fuel as you want, that's going to be a great thing."

Pre-Pook qualifying consisted of two 30-minute sessions, with the field split in half, on Saturday. This year there is one 45-minute session with all cars running—but each allowed only 15 laps at speed—on Friday and another Saturday. Low qualifier for each session receives one series point.

"This is much fairer," da Matta

Da Matta

none to beat. That is the downside. On the other hand, I think Gil and Helio lost a lot more by switching to the IRL than we did by losing them.

"The [class of] drivers in the IRL is not even close, and I am sure before the year is out that both Gil and Helio will miss road racing. That's how they grew up racing in Brazil. Now they're only on ovals. I would think it would get awfully old."

Da Matta's highest points finish was fifth last year, but championships are not new to him. He won Indy Lights in 1998, including a win at Long Beach, and before coming to the U.S., won six Brazilian titles in Formula 3, Formula Ford and karts.

If he wins Sunday, da Matta will equal the CART record for consecutive wins at four. Al Unser Jr. did it in 1990 and Zanardi in 1998.

Although Da Matta has won three times on ovals (Chicago 2000 and Mexico 2001, 2002), and on a

Gentilozzi Shows He's Game on Last-Lap Jam

By MARTIN HENDERSON
TIMES STAFF WRITER

For anyone enjoying late-race theatrics, it was hard to top the Johnson Controls 100 in the Trans-Am Series for the BF Goodrich Tires Cup.

Three drivers led the last lap, the leader going into the last lap made contact with other drivers at least four times in the final minutes, and **Paul Gentilozzi** went from third to first with only three turns remaining to score the 24th victory of his career.

The timed race was scheduled for 51 laps or 70 minutes, and it went 45 laps. The wild finish began with **Boris Said** taking the white flag and second-place Gentilozzi and **Justin Bell** trailing. Said went into Turn 1 too hot, and Gentilozzi closed the deficit—bumping the back of Said's Panoz Esperante.

They continued to battle to the back straightaway on the 1.97-mile layout on the Long Beach streets, their cars touching as Said tried to block the three-time champion.

As they tangled and Gentilozzi backed off, Bell—who won last year's season finale—made a desperate move past Said, "but I think he forgot to brake," Gentilozzi said. Bell appeared to take the lead ever so briefly before making contact with Said, and in Turn 8, he spun nearly 180 degrees as Gentilozzi moved into first on the 11-turn course.

"That last straightaway was Beach for the fourth time. Said, who started second and did not take tires, was second and **Johnny Miller**, also in a Jaguar, was third. Bell, who started 15th in a Corvette, finished eighth.

The climactic end was set up by a caution flag with nine laps to go, wiping out about a seven-second lead that Said had built by pitting early and then passing last year's race winner, **Lou Gigliotti**, as the leaders pitted on Lap 24.

"I hated to see that last yellow come out because we had this race won easily," Said said. "No way they were going to catch me."

Canada's **Michael Valiante** made a daring pass of **Alex Gurney** on Lap 20 and scored his first victory in the series, giving Lynx Racing its second consecutive victory at Long Beach in the Toyota Atlantic Chevron Challenge.

The timed race was shortened from 32 to 29 laps because it was slowed by 14 laps under caution. A first-turn accident on the first lap brought 24 of the 27 cars to a standstill. Gurney, Valiante and the man who caused the pileup, third-place finisher **Jon Fogarty**, were the only drivers to get through the first turn.

Fogarty made an aggressive attempt at the lead from the inside of the second row, but locked his wheels and slid past the apex of the turn, sending pole-sitter **Joey Hand** and No. 2 qualifier **Ryan Hunter-Reay** into a tire wall.

ney protecting the inside. Valiante made a gutsy pass on the outside going into the first turn.

"I knew that was probably going to be my last chance," Valiante said. "I couldn't come off the hairpin [at the front of the straightaway] as good as him, but I was running a little less downforce. I don't think I would have passed him if I didn't pass him right then."

Valiante wobbled through the turn, coming within inches of the wall, and was never challenged, winning by 2.812 seconds.

"I pulled up a little because I thought he wasn't going to make it," said Gurney, who lives in nearby Newport Beach. "When he made it, I said, 'You lucky dog.'"

In seven Atlantic races, Valiante has never finished lower than sixth.

Fogarty retained the series lead with 35 points, three more than Valiante. **Rocky Moran Jr.**, who finished fifth, has 21 points, followed by **Luis Diaz** with 19 and Gurney with 17.

Hunter-Reay, who finished 18th, broke the Atlantic course record with a lap of 1 minute 18.456 seconds, and was more than a second faster than anyone else in the field.

"I feel like the entire Atlantic field got lucky today because I had the best car I have ever driven," he said.

It was particularly heartbreaking to Dorricott Racing's Gurney, who started fourth, because his father, **Dan Gurney**—who cele-

ADAM PRETTY / Getty Images

San Luis Obispo's Townsend Bell, 26, is a CART rookie and the series' only young U.S. prospect.

AMERICAN DREAM

Bell Followed His Heart All Way to CART Circuit

By SHAV GLICK
TIMES STAFF WRITER

Between Townsend Bell's junior

Career Highlights

Townsend Bell is one of two

as a third driver.

"It was pretty hectic," Bell said. "I landed on Sept. 11, which was traumatic itself, and then qualify-

Engine Quits, but He's Not Off Track

First one out is not a good thing. Tony Kanaan kicked a retaining wall and put his helmeted head in his hands. Kanaan stared at his No. 10 Pioneer-WorldCom/Mo Nunn Racing Honda-Reynard. The car's engine had died, just stopped. The fast car stood still and Kanaan had just become the first driver out of the Toyota Grand Prix of Long Beach on Sunday.

In two events this season, Kanaan has scored zero points.

For a moment Kanaan's spirits were low. But not for long. Kanaan, a 27-year-old Brazilian whose father had raced stock cars on a Brazilian circuit Kanaan says "is as big as NASCAR," has an understanding of his life and the place racing has in it. He walked off the track waving and smiling at the fans. On the way to his garage, Kanaan signed autographs and told those around him who thought the day a disaster that they didn't understand disaster.

"I have had three big things, very bad, affect my life and myself deeply," Kanaan said. "From these things I have grown in my understanding of what is important and what is not. Being rich and famous and winning all the races would be nice but not so important. Helping my friends and family and being a good person who is happy, that is important."

Last September, in the first major worldwide sporting event held after the Sept. 11 terrorist attacks, Kanaan watched in shock and despair as his best friend, Alex Zanardi, escaped with his life but without his legs after a horrific crash at the American Memorial 500. The race in Germany had been renamed to honor the U.S.

Zanardi, an effusively happy and popular Italian CART driver, had befriended Kanaan nearly a decade earlier. Kanaan had come to Italy with no ability to speak Italian and no money. The teenager slept on a mattress on the floor of the garage of the Italian team that had tabbed Kanaan as a future star. Zanardi, already established as a driver, took it upon himself to help Kanaan learn the language of both the word and the car.

The two had remained friends, close enough to talk on the phone several times a week as Kanaan made his way from Italy and Formula 3 racing; to the U.S., and the Indy Lights series (in 1997 Kanaan won the Indy Lights season championship); and onto the CART series in 1998. Last year, for the first time, Kanaan and Zanardi became teammates on the Mo Nunn Racing Team on the CART circuit.

"It was the best thing, to have my friend on the team," Kanaan said. "And then came the accident."

Kanaan spent six days in the hospital with Zanardi, helping Zanardi's wife, waiting for his mentor to come out of a coma and then come to grips with the new life ahead, the one where Zanardi would be without his legs and without his life's love, racing.

"You want me to say what I have learned from Alex?" Kanaan said. "You must never despair. Alex has never been in despair. He walks now, on his new legs. He buys shoes. He takes care of his wife and his son. We talk on the phone, just as before. No different. I don't treat Alex different because Alex wouldn't want that. We talk like normal and that's how life must be. You make the new normal."

Kanaan is pretty good at making the new normal. When he was 13, a driver of go-karts with a derring-do attitude he inherited from his father, Kanaan suffered the first of the three tragedies of his life. His father, Tony, died of liver cancer, a shocking illness in a man who never drank and always had been strong. Kanaan said that as his father got sicker, "my father put the business of our family into the name of my uncle, his brother.

"Suddenly my mother, my sister and I had no money. We had been comfortable and now we had nothing."

As the man of the little family in Sao Paolo, Kanaan went to work. He gave go-kart driving lessons and built go-karts to sell. Four years later the great Brazilian racer Ayrton Senna, who died in a 1994 race crash, recommended Kanaan to an Italian racing team looking for young, talented Brazilian drivers. "I knew nothing of Italy or the language, but I wanted to race and so I went," Kanaan said. "I had nothing but my love of racing and determination to make it."

In so many senses, Kanaan has made it. His two proudest accomplishments on the track were that Indy Lights title and winning the 1999 Michigan 500. He lives in Miami now with his 23-year-old sister, Karen, who is going to school. He is able to send money to his mother.

But as happy a moment as that 1999 victory was to Kanaan, it is a year that also brought Kanaan deep sadness. Kanaan's other close friend in racing was Greg Moore, the young Canadian who was killed in a 1999 crash in the Marlboro 500 at the California Speedway in Fontana.

"To lose my father and to have the bad consequences for our family, and then to lose Greg, my good, dear friend, that was very hard. From those things, and from Alex's accident, I think I understand life more. I understand the things that matter are having great family and great friends and being there for them.

"People laugh when they hear I live with my sister. But my sister and I were separated for so many years and now we are being reunited in a way. We have so much fun together. I would be the sad one if she moved out."

Kanaan had been pleased with earning the sixth spot on Sunday's starting grid and, on the first turn, had moved up to fifth. His car had recorded the second-fastest practice time Sunday morning. To have the engine quit was a huge disappointment to the team but Kanaan chose to look ahead.

"Tough luck, huh?" Kanaan said. "We'll get 'em next time. That's the only way to respond to something like this because it's totally out of your control."

It is the way of racing and of life. Kanaan gets it.

Diane Pucin can be reached at diane.pucin@latimes.com.

The CART FedEx Championship Series Hits the Beach.
THE 28TH ANNUAL TOYOTA GRAND PRIX OF LONG BEACH
LIVE QUALIFYING ON SPEED CHANNEL
Today & Tomorrow 1:30PM
CART FRIDAY NIGHT ON SPEED CHANNEL
Tonight 7:30PM
LIVE RACE COVERAGE ON FOX
Sunday 12:30PM
SPEED
CHANNEL
NO ONE GETS YOU CLOSER™
CART
To get SPEED CHANNEL call 1-888-22-SPEED. Or call your local cable operator or satellite provider today.
For complete programming information log on to speedtv.com

FedEx
Express

2003

The 2003 Toyota Grand Prix of Long Beach, race three of the CART season, was held on April 13, 2003 in front of a crowd of 95,000.

The Automobile Club Lifestyle Expo had, among other things, a display of the Historic Formula One cars that will race on Sunday morning. One of the historic racecars will be awarded the Pete Lyons Cup for best evoking the memories of 1976-1983 at Long Beach.

Friday was practice and qualifying for all. Saturday featured the Toyota Celebrity Race at noon, the Argent Toyota Atlantic Race at 4:30 PM and then the Rock- N - Roar concert with the "Gin Blossoms".

Sunday, race day, had a flyover by F-18's, the Historic Formula One cars racing at 9:40 AM, the Argent Toyota Atlantic Race at 11:25 AM followed by the popular driver introduction laps in Toyota Tundras, the CART Race at 1:00 PM and the Trans Am Race at 3:45 PM..

The CART race was live on the Speed Channel with the lowest ratings to date, 69,000 watchers, with reporters Scott Pruett, Tom Kendall, Derek Daly and Calvin Fish. Bob Varsha was the lead announcer.

Stars and Cars Are Revving Up for the Long Beach Grand Prix

The city is expecting 250,000 people for the celebrity-studded racing event this weekend — and hoping they will bring $40 million into the city.

By Nancy Wride
Times Staff Writer

A whiff of celebrity and exhaust has already blown into Long Beach, as it does every April with the arrival of the civic cash cow called the Grand Prix.

Actors and athletes, a model and an astronaut — even professional race car drivers — have already screamed around the 1.9-mile course looping the waterfront, prepping for the annual Toyota Grand Prix of Long Beach next weekend.

As many as 250,000 racing fans will crowd the streets of downtown between Friday and Sunday, having bought out every major hotel and high-rise parking lot for a street race advertised as "the world's fastest beach party."

"It's definitely the crown jewel of the circuit," said driver Jimmy Vasser, last year's champion, as he chauffeured yet another reporter around the track at 120 mph.

This was last week, during another Grand Prix tradition: celebrity media day. And with ride-alongs like this, and star presence uncommon to Long Beach — except occasional sightings of native Cameron Diaz — the odds for good press are excellent.

And that is good for the state's fifth-largest city, the size of Cleveland but ever in the

The race is international, with only two American drivers. It is televised and attracts international media: One of Mexico's most popular pro athletes, driver Adrian Fernandez, is a big draw.

This year's race also comes as Long Beach finally appears closer to finishing an ambitious, though controversial, waterfront compound of restaurants and theaters — a project it hopes will help anchor its downtown as a bigger business and tourist draw.

With a backdrop of towering steel and skyscraping cranes, the race course winds around what has been a shifting landscape of massive construction downtown of about 3,000 residential units, plus commercial buildings.

But the construction has not disrupted the race, beyond forcing the relocation of some of the huge grandstands that line Shoreline Drive and other spots.

The Long Beach Convention and Visitors Bureau and the city did not have independent numbers, but use the Grand Prix's estimate that the race funnels approximately $40 million into the city, directly and indirectly, from the private race promoter's fees to corporate sponsorships and tourism. Of the roughly 4,000 hotel rooms in the city, virtually all are sold out at

Luis Sinco *Los Angeles Times*

PRACTICE RUN: *Drivers who will participate in the Long Beach Grand Prix test the course, which winds along city streets. The event draws race-car drivers from all over the world.*

weekend. John Morris, owner of Mum's restaurant on Pine Avenue, in the heart of downtown, said the race draws so much business that even his patio is booked a year out.

"It is the great people-watching weekend," said Morris. "I'm not too sure how many of these people come for the race. I think it's a happening, and Long Beach has been doing it for so many years that people know it's a party. The atmosphere is awesome. There are a lot of celebrities — Paul Newman and David Letterman and all the celebrity drivers."

And, he added, "It couldn't come at a better time for business: economy, the war, the convention business hasn't been real strong. ... So this is something we've been looking forward to. It's three weekends of business in one."

For a city warned by a financial consultant last month to slash workforce costs or face potential bankruptcy, this giant street party is a very good thing.

Crowd control and safety will be especially tricky this year, in light of the war. And the port complex just across Long Beach Harbor from the Grand Prix has been ranked as high as third on the West Coast for potential terrorism targets.

Long Beach Police Sgt. Ron ment's policing at the Grand Prix, said enough officers will be working to manage the crowd. He said last year's attendance was down some, probably a reaction to Sept. 11, but his department is bracing for as many as 250,000 between Friday and Sunday. The largest single-day attendance is Sunday, when promoters estimate 100,000 may attend.

When the race was introduced 29 years ago, Long Beach was a starkly different place.

That first year, race banners blotted out signs for X-rated movie theaters, tattoo joints and porn parlors, and race cars zoomed around a once-great visitors flocked to the shore and the legendary Pike amusement park. By the 1970s, the area had gone seedy.

At that time, a guy who owned a small travel agency cooked up the idea of staging a car race to attract people to Long Beach. His name was Chris Pook, and locally he is something of a legend, if not widely known outside race circles. Pook still divides his time between Long Beach and Indianapolis, where he runs Championship Auto Racing Teams, the NBA of car racing.

"We did have a crazy idea back in the 1970s of what we could do and should do," Pook said last week.

Pook was honored with a plaque, embedded in a monument honoring the Grand Prix, that thanks him for crafting a race that "immeasurably changed the face of an entire city."

Whatever the exact numbers, the Grand Prix has brought enough good in the past generation that even residents who detest the deafening engines and view them as amplified buzzing flies tolerate clogged downtown streets and overflow traffic.

And those who live in the thunder of the race course can be bused, compliments of the Grand Prix, for two-day trips to the zoo and outlet shopping.

Apart from race refugees, the only creatures who do not have it so good are the fish at the Aquarium of the Pacific, which shuts down for four days in April while the race cars

Jourdain's first pole comes at Long Beach

Michel Jourdain qualified on the pole but second fastest Tracy won the race and $100,000 followed by Adrian Fernandez and Bruno Junueira. Tracy got the jump at the start and led for 26 laps, Jourdain took over until lap 56 where Servia jumped into the lead. He was replaced briefly by Fernandez, then Bourdais, then took the lead again on lap 66. His car stalled in the pits and Tracy passed him on lap 84 and went to on to win. There were three cautions; Lemaria stalled on course, Yoong hit the wall and Lavin and Moreno ran into each other.

Tracy, Fernandez and Junqueira

Peter Reckell won the Toyota Celebrity Race followed by Jesse James and Adam Carolla, all amateurs. The first professional was Jeremy McGrath in fourth. Boris said won the Trans Am Race with Johnny Miller second and Scott Pruett third. Townsend Bell won the Indy Lights race with Dan Wheldon second and Derek Higgins third. A.J. Allmendinger

won the Atlantic Race with Aaron Justus second and Jonathan Macri third. The Historic Formula One race was won by Danny Baker in an M23 McLaren, followed by Charles Nearburg in a Williams and Erich Joiner in another Williams.

(See all the race results on the attached DVD)

Tracy's Sitting on Cloud Three

He takes advantage of Jourdain's trouble to win Long Beach Grand Prix for third time. It's also his third straight CART victory.

By SHAV GLICK
Times Staff Writer

Less than 14 miles from his first win as a CART champ car driver, Michel Jourdain Jr.

Michel Jourdain

Roberto Moreno

Paul Tracy

Adrian Hernandez

Tiago Monteiro

Mario Haberfeld

TOYOTA
TOYOTA
TOYOTA
TOYOTA
TELMEX

TOYOTA
TOYOTA
TOYOTA
TOYOTA
TOYOTA
TOYOTA
TOYOTA
TOYOTA

American Monaco?

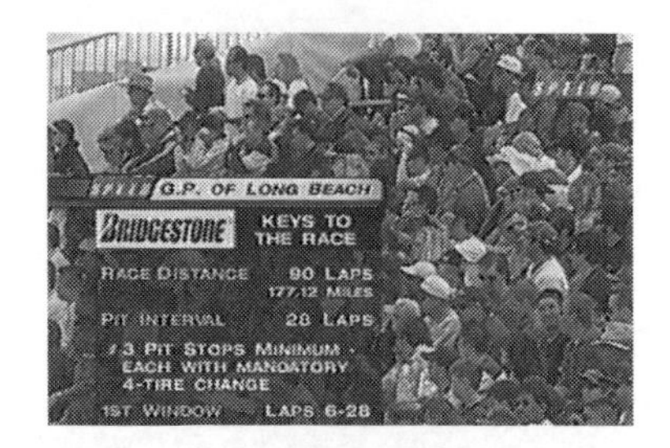

Read the Autoweek article on the attached DVD and watch the video on our youtube channel - www.youtube.com@racinghistoryproject2607

ROCK 'N ROAR

RETURNS!

Southern California's Official Spring Brake returns with a bang!

29TH ANNUAL TOYOTA GRAND PRIX OF LONG BEACH
APRIL 11-13

Five great racing series! Free Saturday musical concert featuring hot-rocking "Gin Blossoms!" Fun activities for the whole family, including:

- Paul Tracy, Adrian Fernandez and the cars and stars of the Bridgestone Presents the Champ Car World Series Powered by Ford
- Historic Grand Prix Race, featuring 24 classic Formula One cars
- Toyota Pro/Celebrity Race, Argent Mortgage Toyota Atlantic & Trans-Am
- X-Games and exhibits in the Automobile Club of Southern California Lifestyle Expo, aerials show and much, much more!

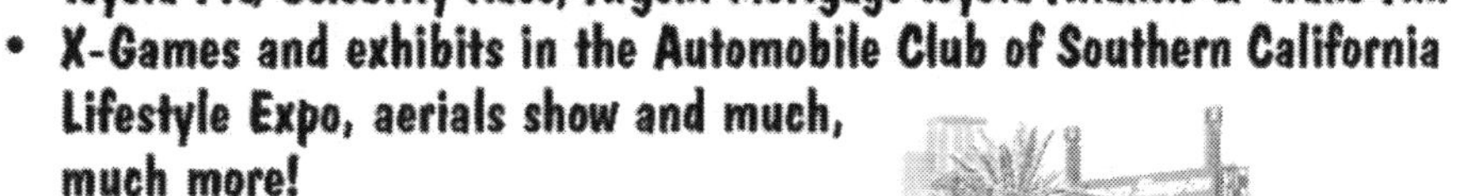

Call now for tickets **888-82-SPEED or order at longbeachgp.com**

2004

The 2004 Toyota Grand Prix of Long Beach, the first race of the “Bridgestone Presents the Champ Car World Series”, replacing the bankrupt CART, was held on Aprll 18, 2004 in front of a crowd of over 78,000. Now adding to the new sanctioning body attraction, a two seater Champ Car for press rides, hew tire and pit stop rules and more horsepower with “push to pass”.

The week began with a 30th Anniversary Party at Harry O’s in Manhattan Beach and Wednesday’s media luncheon; not well attended by drivers as Champ Car had scheduled a test day at Fontana.

Thursday was the race car display on Pine Avenue and the “Tecate Party on Pine”, featuring a rock concert, pit stop demonstrations and a radio show broadcast from Mum’s Restaurant. The Automobile Club Lifestyle Expo opened Friday morning as did practice and qualifying for all. Saturday featured more practice and qualifying and the Toyota Celebrity Race at noon.

Sunday, race day, was sunny and warm and had the Historic Grand Prix Race at 9:45 AM, Formula Atlantic Race at 10:30 AM, driver introduction laps in Toyota Tundras and then the CART Race at 1:00 PM followed by the Trans Am Race at 3:45 PM..

The race was live on XTRA Radio. ABC had the IRL race at Phoenix with Jack Arute, Paul Page and Scott Goodyear while Champ Car had was on Spike, formerly TNN. The broadcast is tape delayed three hours on the west coast. Tom Kendall, Derek Daly and Calvin Fish were on hand as reporters and Bob Jenkins was the lead announcer.

Junqueira nips Tracy for Long Beach provisional pole

BY MIKE HARRIS
ASSOCIATED PRESS

LONG BEACH — Bruno Junqueira got a clear lap and Paul Tracy didn't.

That was the difference Friday. Junqueira, last year's runner-up to Tracy in the Champ Car World Series championship, won the first point of the season

Due to the new Champ Car consolidation, 18 cars started, Bruno Junqueira won the pole but Paul Tracy won the race with Junqueira second and Sebastian Bourdais in third. Tracy jumped into the lead at the start, using his "push to pass" button, putting him in front until lap 58. Junqueira led for a couple of laps during pit stops as did Carpentier. But Tracy reclaimed the lead and won by six seconds over Junqueira. Only eight cars finished on the lead lap. And there was only one big crash, turn one , lap one, knocked out Vasser, Marques and Sperafico. Allmendinger and Phillipe were also involved but were able to continue.

RICHARD HARTOG *Los Angeles Times*

LIFTING OFF: *The back wheels of Alex Sperafico's car rise off the course during the race's only accident, which knocked out Sperafico and two others.*

Tracy Wins His Fourth Long Beach Title

[*Long Beach, from Page D1*]

to-wire, but from the moment he used his boost button to jump into the lead, he was in complete command. He led 78 of the 81 laps in the pale blue Forsythe Lola-Ford that won the championship last year.

The race around the 11-turn, 1.968-mile temporary circuit laid out on the seaside lowlands lasted 1 hour 44 minutes 12.3 seconds, but for all intents it ended seconds after the 18 cars took the green flag on the second lap.

As the front row of Bruno Junqueira and Sebastien Bourdais, Newman-Haas teammates, reached the start-finish line, Tracy jerked his car sideways, hit the boost button that gave an extra 50 horsepower for about six seconds, and shot past the pair from his position in the second row.

Junqueira, the pole-sitter, hit his button too, but not quickly enough to stave off the flying Tracy. By the time the pair reached the first turn about a half-mile into the race, Tracy was gone.

The only time he was out of the lead was on the first lap when officials waved the yellow flag because of a too-ragged lineup and Junqueira was credited with leading, and on the two occasions when Tracy pitted and his Forsythe teammate, Patrick Carpentier, led a lap both times.

"I've always been aggressive from the second they wave the green flag," Tracy said. "It is always important to get a good start and I managed to get one today.

"I used the push-to-pass button right on the line to pick up more speed. It was really a jolt and then I took advantage of Bruno's draft and charged to the inside. I saw that Sebastien was squeezed against the wall, so I had a wide-open lane.

"Bruno braked earlier than I anticipated and because I had so much momentum my car started to slide and got sideways. By the time we got to the first turn, though, I had the lead and just pushed as hard as I could to pull away."

Junqueira, who finished second to Tracy in the championship race last year, was philosophical about seeing Tracy go by.

"I knew that Paul was going to be pushing at the start," Brazil's Junqueira said. "He has always done that, only he used to crash. Now he doesn't crash and he wins races."

Junqueira finished second, 5.68 seconds back, with teammate Bourdais a close third. No one else was within 15 seconds of the winner, although eight cars finished on the lead lap. Tracy's average speed of 91.785 mph was a record, bettering his year-old speed of 91.590.

The new rules involving the push-to-pass button and the mandatory use of soft tires after one pit stop proved interesting to fans who could watch the drama of decision-making on monitors. Each driver was given 60 seconds worth of added boost, which he could use at any time or in any amount.

Seven drivers used their limit and most used nearly all of theirs, except Tracy, who did not need any more help after the six or seven seconds he used at first.

"Once I got in front, I tried to build up a good lead and hold it before we pitted, then when I got back out I did the same thing," Tracy said. "We didn't use the soft tires until our last stop when we had only 21 laps left. We were concerned about how they would last."

Apparently the concern was unjustified. His 79th lap, only two from the end, was his fastest. Six other drivers had faster single laps than Tracy, but the consistency of the champion paid off.

The only accident of the race, remarkable for the first time out for teams that had not raced in seven months, came on the first turn when four or five cars tried to cram through a lane big enough for no more than two. Knocked out of the race almost before it started were former winner Jimmy Vasser, Tarso Marques and rookie Alex Sperafico. Two others, rookies A.J. Allmendinger and 17-year-old Nelson Philippe, were involved but regrouped and continued.

"First corner, first race of the new season," an exasperated Vasser said. "I don't really know what to say. I got hit in the left rear tire. Then there was chaos with guys weaving all over the place. We just got spun around and collected."

Philippe, who went on to finish 13th, became the youngest starter in CART-Champ Car history. Michel Jourdain Jr., who was 19 in his first race in 1996, had been the youngest.

"I feel amazing," said Philippe, a long-haired Frenchman who drove in the Barber Dodge series last year. "I did everything right, focused on the track and my drive.

"I got out of the tangle early in the race and paid attention to let the leaders pass."

The champ cars' next race is May 23 in Monterrey, Mexico.

Max Papis won the rainy Toyota Celebrity race and was the first professional followed by the first amateur, Chris McDonald and Andrew Firestone. The Toyota Atlantic Race was won by Ryan Dalziel with Andrew Ranger second and Bryan Sellers in third. Paul Gentilozzi won the Trans Am followed by Greg Pickett and Boris Said.

(See all the race results on the attached DVD)

Catching a Ride in the Fast Lane

By Martin Henderson
Times Staff Writer

LELAND HILL

TAKING A SPIN: *Champ Car driver Jimmy Vasser shows Times writer Martin Henderson the course.*

The closer I got, the scarier it looked. I had prepared for this, with two nights of anxiety and a day of avoiding food. The car, 16 feet long, was sleek, its lines reminiscent of the Champ Cars that had just qualified for today's 30th Toyota Grand Prix of Long Beach.

Except this silver streak of a car was deformed. There was a hump, like a sleek thoroughbred camel, behind the driver. In it was a space carved out for a passenger.

They call these single-seaters for a reason, but this was a Champ Car two-seater. One car, one driver, one rider.

My driver was Jimmy Vasser, who won this race in 1996 along with the series championship. I told him he had been my favorite driver "for about 20 minutes." Vasser laughed. I reminded him about a pace car ride he had given me a few years ago. Approaching the back side of the course, I recalled, "You had your hand down in the floorboard, fumbling around to pick up a water bottle, doing about 75 mph."

"I remember that," Vasser said. "We're going to go a hell of a lot faster today."

Those were Jimmy's last words to me before he climbed into the cockpit and I put my life in his hands.

"Warm it up," he was told.

The engine started. A gentle rhythm, not unlike a hotel massage bed. Vasser hit the throttle in spurts. The vibration shot up my spine.

There have been other rides that gave me a lot of vibration. A GT car at Del Mar in 1992, a Ferrari on the California Speedway road course in 2002. Those rumbles were more guttural. This was compact. Just like this crevice that doubled as a seat.

I have never been strapped so tight in such little space. Nor so thankful.

The warmup over, I signaled for Russell Cameron, team manager for PKV Racing, the team that is partially owned by Vasser and the one that converted this 2000 Reynard into Champ Car's newest marketing tool.

"If something happens," I asked,

> The wind buffeting the helmet forced my head back over my left shoulder, but I see Vasser leaning into the right-hand bend along Shoreline Drive, reaching 176 mph. We do nearly three times the freeway speed limit, except we reach it much more quickly.

used for today's main event. It's not some motorcycle engine stuffed inside a shell. It's a monster, a real

It's throttle-brake-throttle-brake into Turns 4 and 5, where we swing wide toward the wall. A pro-

hits the throttle like John Force. My bottom never touches the seat on Shoreline. I don't remember breathing. I get six laps. I can't imagine keeping up this pace for 81.

The only word that comes to mind is "amazing." Sure, it was incredible, but the technology, the power, the forces on the body were simply amazing. It reminded me of my youth standing next to railroad tracks as the twice-daily freight train went past, except in a Champ Car, you must imagine standing between two trains.

Vasser jumped out of the cockpit. "That's faster than my car," he told Cameron, the team manager.

Nelson Phillipe

Mario Dominguez

Tracy, Junqueira and Bourdais

Sebastian Bourdais

Oriol Servia

Patrick Carpentier

Jimmy Vasser

MOTORSPORTS

VINTAGE LONG BEACH

Read the Autoweek article on the attached DVD and watch the video on our youtube channel - www.youtube.com@racinghistoryproject2607

Associated Press/Michael Kim

Ryan Hunter-Reay recently drove a 750-horsepower Ford-Cosworth/Lola through Times Square in New York during the filming of a Champ Car commercial. "I love Champ Car, and I love these machines," he said. "They're made to race."

WATCHING THE ROAD AHEAD

Champ Car series has put bankruptcy and the defection of some top teams in the rear-view mirror as season opens today

Strategy to play major role in Champ Car opener

BY MIKE HARRIS
ASSOCIATED PRESS

LONG BEACH — Sebastien Bourdais had the best qualifying lap and the best line Saturday at the Long Beach Grand Prix.

Looking ahead to Sunday's season-opening race, Bourdais, last year's top rookie in the Champ Car World Series, noted the drivers and teams are facing a number of questions about new rules and equipment.

Asked how they were going to deal with rules requiring two green flag pit stops, a mix of hard and soft tires and a new "push to pass" button that will give the turbocharged engines a limited 50-horsepower boost to promote passing, the Frenchman shrugged and shook his head.

"I'm sure the computers are going to smoke this evening," he said, drawing chuckles from Newman/Haas Racing teammate Bruno Junqueira and defending series champion Paul Tracy.

Associated Press

Champ Car World Series driver Sebastien Bourdais *goes around the hairpin turn during the second day of racing in the Grand Prix of Long Beach Saturday. Bourdais had the best qualifying lap of the day.*

Kendall wins pole in return to Trans-Am

LONG BEACH — Four-time series champion Tommy Kendall, returning to Trans-Am after a five-year absence, won his 40th career pole Saturday in qualifying for the season-opening race at the Long Beach Grand Prix.

Driving a Jaguar XKR with a production 4-valve V-8 engine for the Rocketsports team, Kendall complete a lap on the

gamble with his tires pay off for his third career victory.

Crawford pulled away on two restarts in the final 34 laps, then had to do it again after Steve Park hit the wall with just over three laps to go in the NASCAR truck series race.

Because series rules provide for a two-lap, green-flag finish, Crawford had to wait until the 253rd lap to do it again.

Setzer finished .365 seconds

REMINISCENCE

One Man Had Street Smarts

Dismissed three decades ago as 'silly,' 'crazy' and 'nuts,' Pook's vision of a race through downtown Long Beach has become a piece of Americana.

Papis Drives Like a Pro in Victory

By MIKE KUPPER
Times Staff Writer

Max Papis, who less than a year ago was driving in the Champ Car World Series — and wouldn't mind rejoining the open-wheel brigade — showed the amateurs how things should be done Saturday when he won the pro-celebrity race at the Toyota Grand Prix of Long Beach.

Papis, after observing the 30-second handicap required of the pro drivers, threaded his way through the field of 13 celebrity racers, passing actor **Chris McDonald** for the lead on what proved to be the last lap. That left McDonald as the celebrity winner.

The race, with all drivers in identical Toyota liftback sedans, was to have been 10 laps over the 1.968-mile seaside course but was cut a lap short because of emergency equipment on the track after soap actor **Peter Reckell**, last year's winner competing as a pro, had hit the wall in Turn 1.

An unhurt Reckell was only one of several drivers confounded by intermittent rain on an already wet track. Slides into the tire walls were frequent.

Not that the wetness bothered Papis, driving professionally this season in several major sports car series.

"I was just trying to have a very good time with Peter," he said. "I told everyone before the race that the key in the rain was to drive fast but smooth."

Actor **Sean Astin** took good-natured exception to that, saying, "You lie, Max. I was the smoothest guy out there and I came in fourth [among the celebs]."

2005

The 2005 Toyota Grand Prix of Long Beach, round one of the "Bridgestone Presents the Champ Car" season, was held on Aprll 10, 2005 in front of a crowd of over 75,000. This was a critical year with contracts expiring with Toyota and Champ Car - Would either or both renew ? Meanwhile, race team owners Kevin Kalkhoven and Gerald Forsythe purchased the Grand Prix Association of Long Beach, ensuring continued operation.

The Meet The Drivers event was held in the Port of Long Beach with everyone bussed in. The Automobile Club Lifestyle Expo featured the Freestyle MX Tour of motorcycle stunts, driving simulators and go kart races on the roof of the parking garage.

Friday was practice and qualifying for all plus Tecate Fiesta Friday.. Saturday featured the Toyota Celebrity Race at noon and the Rock-N-Roar concert and the Miss Toyota Grand Prix Pageant. New this year was the Long Beach Drifting Challenge.

Sunday, race day, had the Toyota Atlantic Race at 11:25 AM, then the driver introduction laps in Toyota Tundras, the CART Race at 1:00 PM followed by the Trans Am Race at 3:15 PM..

TV coverage on NBC missed the race start due to longer than usual football game. ABC with commentators Paul Page, Scott Pruett, Parker Johnstone and Jon Beekhuis.

Tracy knows his way around

A four-time winner at Long Beach, he wins the pole for today's race.

ASSOCIATED PRESS

LONG BEACH - Paul Tracy spent most of last season watching the Newman/Haas Racing duo of Sebastien Bourdais and Bruno Junqueira win races and listening to people question the commitment of the former series champion and his racing team.

After running away with the Champ Car World Series title in 2003, Tracy won only twice - including the season-opening Toyota Grand Prix of Long Beach - in 2004. He slipped to fourth in the standings as Bourdais and Junqueira combined for nine wins and finished 1-2 in points.

A new season begins today at

Associated Press/Mark J. Terrill

Paul Tracy celebrates after winning the pole for today's Toyota Grand Prix of Long Beach with a track-record lap of 104.982 mph.

"It gets me upset when you pick up some motor sports magazines that say the Forsythe team

League will be made within 30 to 60 days following today's race.

Champ Cars have been at Long

that a Hendrick Motorsports ai plane had crashed nearby, killi all 10 people aboard.

Thoughts of the lost team mer bers and friends are constar Johnson said. He views today a chance to pay tribute with a other winning run.

"I'd love to be able to go ba myself and go through all the ce emonies," said Johnson, one four Hendrick drivers. "Eith way, I just hope it's a Hendri car so we can keep the spi alive."

Craftsman Trucks in Martinville, Va. - Bobby Labonte w the Kroger 250 and became t first driver to win in each NASCAR's top three series at Ma tinsville Speedway.

Labonte passed Chad Chaff for the lead with 30 laps to g then held off Ricky Craven ar Ron Hornaday on a restart wi five laps to go. His Chevrolet be

Paul Tracy had the pole but Sebastian Bourdais won, the first of his three wins, followed by Bruno Junqueira and Mario Dominguez. The lead changed twelve times with Tracy and Dominguez swapping the lead until Tracy bobbled in turn 11, giving the lead to Junquiera. Bourdais took over for a couple of laps on lap 30. Vasser, Tracy, Wilson and Bremer each led for a few laps until Bourdais took over for good on lap 70, winning by almost five seconds over Junquiera..

There were four cautions caused by; Phillipe on lap 8, del Monte on lap 32, Sperafico on lap 45 and Ranger on lap 72.

(See all the race results on the attached DVD)

Bourdais cruises to first victory at Long Beach

He's never seriously threatened after passing two-time defending champ Tracy on 38th lap

By Mike Harris
AP motorsports writer

LONG BEACH — Sebastien Bourdais was amazed when he drove past Paul Tracy for the lead with relative ease in Sunday's Toyota Grand Prix of Long Beach.

Bourdais, coming out of the pits two laps after Tracy's first stop, noted that passing Tracy anytime is "usually a handful," but on the streets of Long Beach, where Tracy had won the last two races and four overall, and with cold tires, "it's really tough.

"My crew was surprised I was able to get by PT like that and I was, too. I came on the radio and said, 'How about that!' It was really surprising."

Tracy, who finished second, shrugged off Bourdais' strong move on the 38th of 81 laps, saying, "As you can image, my reaction was the opposite."

But Tracy pointed out he was just going along with the new Champ Car World Series edict prohibiting blocking.

"ing," Tracy explained. "In the past you'd see guys weaving and moving and, if you tried to pass for the lead, you'd wind up crashing. Today, we had three clean passes for the lead and that's good for the sport."

Tracy tried hard to make a race of it after the pass, but this one belonged to Bourdais, the reigning series champion, who gave his Newman/Haas Racing teams its first Long Beach victory in 19 years.

The last time the team co-owned by Chicago businessman Carl Haas and actor Paul Newman won America's longest running downtown street race was in 1987, when now-retired Mario Andretti finished first and Bourdais was 8 years old.

And it could be the last time the Champ Cars, which replaced Formula One here in 1984, race at Long Beach. This is the final year of the contract and Long Beach officials are considering replacing the Champ Cars with the rival Indy Racing League in 2006.

Bourdais started fourth, but came ond to Tracy. Once he got past the Canadian driver, Bourdais built a pair of leads of more than 5.5 seconds, both erased by full-course caution flags. Tracy was unable to take advantage of those opportunities.

The two leaders made their final stops on Lap 61 and Bourdais easily beat Tracy's Forsythe Racing entry back onto the track. After Jimmy Vasser and rookie Andrew Ranger, both driving a different pit strategy, made their final stops on Lap 70, that put Bourdais and Tracy back on top, with the Frenchman leading by 7.6 seconds.

When rookie Bjorn Wirdheim bumped Ranger into a tire wall on Lap 72, bringing out the fourth and final caution of the day, Tracy had one more shot at the leader. But it was no contest.

He got held up slightly by the lapped car of rookie Marcus Marshall, who drove into the pits moments before the restart. By the time Tracy got up to speed, Bourdais was long gone.

Bourdais, who won seven of 14 races last year, took the green flag on Lap 76 and, while Tracy briefly fought off a challenge for second by Bruno Junqueira, the other Newman/Haas

Associated Press/Matt Sayles

Sebastien Bourdais speeds around the street circuit to win the season-opening Grand Prix of Long Beach in the Champ Car World Series.

Katherine Legge won the Toyota Atlantic Race, the first win by a woman, with Antoine Bessette in second and Charles Zwolsman in third. Rhys Millen won the Toyota Celebrity Race and the professional class with Ryan Arciero second and the first amateur, Frankie Muniz, in third. Randy Ruhlman won the Trans Am Race followed by Greg Pickett and Michael Lewis.

ASSOCIATED PRESS

Katherine Legge talks on the phone to her mom after winning the Toyota Atlantic race in Long Beach on Sunday — becoming the first woman to win a race in the developmental series for the Champ Car World Series.

A Race to the Finish

With key contracts about to expire, the Long Beach Grand Prix faces changes next year.

By Shav Glick
Times Staff Writer

After 30 years as America's [illegible]

Key Changes Are in the Offing

[*Grand Prix, from Page D1*]
tract as title sponsor.

"There is one thing of which there is no doubt, and that is that there will be a race Sunday, and there will be a race next year," said Jim Michaelian, president of the LBGP. "Our contract with the city of Long Beach for a race extends to 2010.

"I think you can be pretty sure that Toyota will be here too. We have too much invested in each other not to stick together. Their pro-celebrity race on Saturday is the gold standard for similar events worldwide."

Just to make sure there is enough for everyone to see and do this weekend, Michaelian has included exhibitions of drifting, extreme motorcycle jumping, participatory kart racing and rock concerts.

The speculation is about the main event next year.

"I hate to hear talk about 2006 and beyond because it detracts from the job at hand, and that's to put on the best open-wheel, open-cockpit race in the world on Sunday," Michaelian said. "Champ Car came through under worse conditions last year, after CART collapsed, so let's see what they do this time." Nevertheless, speculation persists, and even Michaelian says there are ways in which Champ Car needs to change.

"First off, their season starts way too late," he said, his race being Champ Car's opener. "Fans need to become emotionally involved, and to do that they need someone to follow. With no [previous] races, there are no Champ Car drivers to follow. By the time we open the season, NASCAR will have had five [Nextel] Cup races, and the IRL, the NHRA and both sports car organizations at least three races each.

"We're like an island in the schedule. . . . After Sunday, it's five weeks to the next race in Mexico and two more weeks before they race again in the U.S. How can you build up any interest with continuity like that? How can fans identify with new drivers if they aren't racing?"

Among Champ Car's new drivers are Bjorn Wirdheim of Sweden, Ronnie Bremer of Denmark, Marcus Marshall of Australia, Timo Glock of Germany and Andrew Ranger of Canada, all of [illegible]

'If an opportunity presents itself to sit down and have a discussion after this year's race, we would welcome it.'

Adam Saal, Grand American director of communications, on the possibility of becoming part of the Long Beach Grand Prix

queira and A.J. Allmendinger. Michaelian said that at least 18 cars would start.

Michaelian, 62, has been part of the Long Beach race since the day Chris Pook startled the sporting world with the announcement that he would promote a race through the streets of downtown Long Beach in 1975.

"When I heard that some Englishman with a travel agency was going to put on a race, I called Pook and told him there was no way that there could be a race in my town without me being part of it," said Michaelian, a 1964 graduate of UCLA and a racing enthusiast. "He hired me officially as controller, but there were only five or six of us in the operation and we all did what had to be done."

The Grand Prix staff now numbers 25 full time, with another 25 part timers for the race.

Pook, who ran the Grand Prix until he resigned in 2002 in an unsuccessful attempt to save CART, often said that, "I had the dream and the vision, but Michaelian is the glue who held it together. Without him I don't know if we would have made it."

Pook is no longer involved with the Grand Prix Assn., although he remains part of the Long Beach scene, having been awarded a $40,000 consultant's job for the city's annual International Sea Festival.

"I had grown up in Long Beach and I had a passion for racing, so it seemed like the perfect setup for me," Michaelian said. "I've been blessed to have an opportunity to do what I love in a business climate in a city I love. I just want to keep that going."

no such compunctions.

There is always the possibility that Champ Car will remain. If that doesn't happen, though, there is the rival Indy Racing League. Some say that the IRL street race last Sunday in St. Petersburg, Fla., the first non-oval race in IRL history, was really a dry run to prepare for Long Beach. IRL founder Tony George has always said that he would like to fit a good street race into his oval schedule, and there's no better street race than Long Beach.

It doesn't hurt, either, that Toyota has a strong presence in the IRL as an engine manufacturer.

Then there are the sports car organizations, Grand American and the American LeMans. Both put on exciting races, such as the one last Sunday at California Speedway where 45 cars started the Grand American race.

NASCAR? Outrageous as it might sound, the idea of Jeff Gordon, Dale Earnhardt Jr. & co. thundering down Shoreline Drive in their Nextel Cup cars is no more outlandish than was the thought of Formula Ones screaming down Ocean Boulevard in 1975.

A more likely NASCAR scenario would be a truck race. Toyota is part of the Craftsman Truck series, and the trucks have a history of racing on road and street courses.

Adam Saal, Grand American director of communications, probably could have been speaking for them all when he said, "I can't think of any series that wouldn't want to be part of the premier street race in the country, but I want to make it clear we have not had any official discussions with the Long Beach organizers and respect the fact that their main area of focus is on [this weekend's] Grand Prix. If an opportunity presents itself to sit down and have a discussion after this year's race, we would welcome it."

As the longtime sponsor, Toyota is sure to have some say in what happens.

"We'll wait to hear back from Dover, and listen to what Jim Michaelian has to say, and then we'll sit down and see what they decide," said Les Unger, Toyota motorsports marketing director. "As sponsors we might make a recom- [illegible]

BOURDAIS WINS THE BATTLE ON THE BEACH

Read the Autoweek article on the attached DVD and watch the video on our you tube channel - www.youtube.com@racinghistoryproject2607

NOTES

Steadiness Pays Off for Millen and Muniz

By MARTIN HENDERSON
Times Staff Writer

If ever there was a Toyota Pro/Celebrity race that should have been a foregone conclusion, it was Saturday's at the Toyota Grand Prix of Long Beach.

Pole-sitter **Ingo Rademacher** not only qualified faster than his celebrity peers, he was even faster than the professionals. And Rademacher knew it.

"We were sitting there [in a drivers

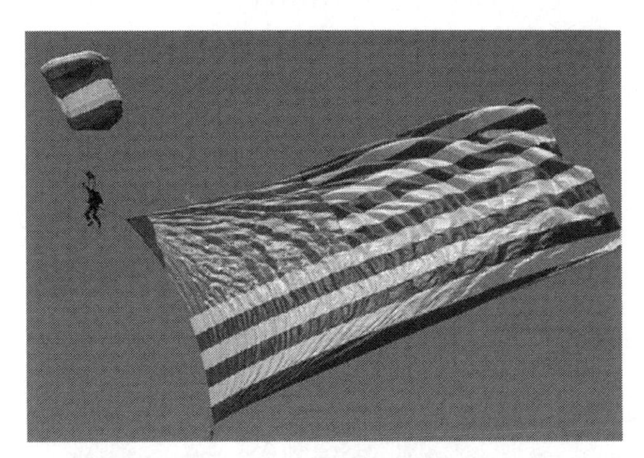

Justin Wilson

Marcus Marshall

Sebastian Bourdais

Bruno Junquiera

Timo Glock

Ryan Hunter Reay

2006

The 2006 Toyota Grand Prix of Long Beach, round one of the "Bridgestone Presents the Champ Car World Series" season, was held on Aprll 9, 2006 in front of a crowd of over 80,000.

The Automobile Club Lifestyle Expo featured the "Freestyle MX Tour" of motorcycle stunts, driving simulators and go kart races on the roof of the parking garage.

Friday was practice and qualifying for all plus the Tecate Fiesta Friday. A busy Saturday featured the IMSA Crown Royal Rolex Grand Am and the Toyota Celebrity Race at noon and the "Rock-N-Roar" concert in the evening. Sunday, race day, had the Champ Car Mazda Atlantic Race at 10:35 AM, the CART Race at 1:00 PM. Formula Drift at 3:20 PM and the Speed GT Challenge at 4:00 PM..

The Long Beach Motorsports Hall of Fame, new this year, was dedicated by Long Beach Mayor Beverly O'Neill with two inaugural members; Dan Gurney and Phil Hill.

TV coverage was on NBC with commentators Derek Daly and Jon Beekhuis.

Bourdais will be on pole at Long Beach Grand Prix

The Associated Press

Sebastien Bourdais shows no signs of slowing down.

After two straight Champ Car World Series titles, the 27-year-old Frenchman appears just as motivated as the first day he drove one of the 750-horsepower cars in 2003.

On Saturday, the defending champion in the Toyota Grand Prix of Long Beach continued his success on the downtown street circuit, backing up his provisional pole from the previous day by setting a track record in winning the 19th pole of his Champ Car career.

AUTO RACING

ing the Indianapolis 500.

Pruett, Diaz win in Ganassi car: Scott Pruett and Luis Diaz co-drove Chip Ganassi Racing's No. 01 Lexus Riley to victory in the inaugural Grand American Rolex Sports Car Series race at Long Beach.

After starting from the pole position in the 90-minute timed event, Diaz surrendered the lead to Patrick Long in the No. 23 Alex Job Racing/Emory Motorsports Porsche Crawford on the sixth of 52 laps. Diaz

en Busch races and led all but 55 laps this season. Busch, making his debut in the No. 39 Penske Dodge that Ryan Newman also drives, is the seventh different Busch winner this season.

. Busch celebrated with the obligatory burnout along the frontstretch before getting out of his car. After an awkward head-first dive into the infield grass, he rolled over and did "snow angels" in the bright, warm sunshine. It was similar to his celebration on the finish line at Bristol two weekends ago when he won the Cup race at the track plagued by winter weather.

MARK J. TERRILL *Associated Press*

PROVISIONAL POLE: *Sebastien Bourdais said he was used to struggling on the first day at Long Beach, but he gets through the first turn en route to the fastest first-round qualifying lap of 104.689 mph.*

Sebastian Bourdais had the pole and won the race, setting a new lap record and race record,, the second of his three wins. Second was Justin Wilson, 14 seconds behind, with Alex Tagliani in third. Bourdais led a total of 70 laps with Jan Heylen leading four. Bourdais was the benefit of a first lap crash behind him that took out four contenders. Dominguez hit Tracy who was then hit by Servia, Junqueira and Allmendinger. Bourdais picked up $85.000 for the win.

Bourdais Cruises as Others Crash

The pole-sitter is out in front of a five-car wreck on first turn and easily wins Grand Prix of Long Beach.

By JIM PELTZ
Times Staff Writer

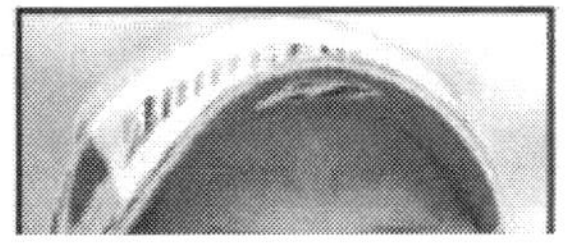

sessions as well — that the wreck's impact might have been moot.

SPENCER WEINER *Los Angeles Times*

BIG GULP: *Sebastien Bourdais takes a drink after winning the Toyota Grand Prix. "The car was just awesome today," he said.*

Scott Pruett and Luis Diaz won the inaugural IMSA Crown Royal Grand American Race in a Ganassi Lexus Riley followed by the Rockenfeller / Long Porsche and Shane Lewis / Bill Auberlen BMW. In the Toyota Celebrity race, Bucky Lasek won with John Elway second and Martina Navritilova third. The first professional, Todd Bodine, was fourth. Andreas Wirth won the Champ Car Mazda Atlantic Race followed by Raphael Matos and James Hinchcliffe. Ron Fellows won the Speed World GT Race with Tommy Archer second and Mike McCann third.

(See all the race results on the attached DVD)

'Speed Racer' Helps Out

A deal involving the cartoon character may give driver Andrew Ranger more races.

By MARTIN HENDERSON
Times Staff Writer

Champ Car driver Andrew Ranger was in a desperate spot late last week. He had brought enough sponsorship to the Mi-Jack Conquest Racing team that the team's second driver.

"I'm very happy, Eric is very happy, and it will help us to get more sponsors," Ranger said. "I mean, the situation can't be any worse."

In that case, Ranger made the best of a bad situation. He was off the pace all weekend with what was thought to be a chassis issue. By the time it was discovered late Saturday that the problem was in the brakes, the team was behind on the chassis setup.

youngest podium finisher ever. He also had six top-10 finishes and finished 10th in the Champ Car standings, one point ahead of 2002 series champion and former Formula One driver Cristiano Da Matta. And, he did it despite five races among the 13 in which his Lola-Ford Cosworth failed to finish. But it reflects the status of open-wheel racing.

"It's a tough situation for a driver," Ranger said. "The money can do everything. I have a ride with Mi-Jack to do a cou-

MOTORSPORTS | CHAMP CAR AT LONG BEACH

WAITER, WE'LL HAVE THE BOURDAIS

FRENCHMAN DOMINATES SNOOZER OF A CHAMP CAR OPENER

Read the Autoweek article on the attached DVD and watch the video on our youtube channel - www.youtube.com@racinghistoryproject2607

ASSOCIATED PRESS PHOTOS

Burning rubber under the Long Beach Aquarium is just another day at the office for drifters. In the sport — which is gaining popularity — racers slide around corners of the track and use a combination of braking, acceleration and self-defense driving to regain control. Below, Kazu Hayashida, sitting in his car, is one of several Japanese drift stars who now compete in the U.S. league.

Jan Heylen

Andrew Ranger

Charles Zwolsman

Cristiano da Motta

Oriol Servia / Paul Tracy Crash

Formula Drift

IMSA Grand Am

Will Power

Sebastian Bourdais

Antonio Pizzonia

Bruno Junqueira

Toyota Celebrity Race

Bourdais, Wilson and Tagliani

2007

The 2007 Toyota Grand Prix of Long Beach, round two of the The Champ Car World Series, was held on Aprll 15, 2007 in front of a crowd of over 100,000 with total attendance over 200,000 for the weekend. This years Meet The Press event was held at the Museum of Latin American Art.

The Automobile Club Lifestyle Expo featured motorcycle stunts, driving simulators and vendors of performance equipment.

Friday was practice and qualifying for all plus Tecate Fiesta Friday". A busy Saturday featured IMSA Crown Royal Grand Am and the Toyota Celebrity Race at 11:30 AM, drifting demonstrations and the Rock-N-Roar concert with "Taking Back Sunday" at 6:00 PM. Sunday, race day, had the Champ Car Mazda Atlantic Race at 10:35 AM, the CART Race at 1:00 PM. Formula Drift at 3:20 PM and the Speed World Challenge Race at 4:00 PM.

TV coverage was on NBC with commentators Wally Dallenbach Jr. and Jon Beekhuis.

The Long Beach Motorsports Hall of Fame installed three new members in 2007; Brian Redman, Chris Pook and the Newman Haas Racing Team.

Bourdais snags pole, but Power still on

The Facts

- **Site:** Streets of Long Beach.
- **Schedule:** Today, race (Ch. 4, 1 p.m.)
- **Track:** Temporary road course, 1.968 miles, 11 turns.
- **Race distance:** 159.408 miles, 81 laps.
- **Last race:** Will Power became the first Australian to win a race in the Champ Car World Series or its predecessor, CART, leading 38 of the 68 laps en route to the victory at the season-opening Vegas Grand Prix. Power won by about half a mile on the 2.44-mile, 12-turn temporary circuit.
- **Last year:** Sebastien Bourdais easily won for the second consecutive year. He trailed only after a midrace pit stop, leading 70 of 74 laps on the course that threads its way through city streets tucked between downtown and the Pacific Ocean. Paul Tracy, Bruno Junqueira and A.J. Allmendinger didn't make it through the first turn of the race.
- **Fast facts:** Three-time defending series champion Bourdais and heralded rookies Graham Rahal and Simon Pagenaud all failed to finish the race at Las Vegas. Bourdais started 16th after a crash in qualifying, overcame three flat tires in the first 24 laps to reach third place and then hit a concrete barrier on Lap 30 and ended up 13th. . . . Tracy has won this race four times but won't race today. . . . Katherine Legge finished sixth at Las Vegas, the best by a woman on a road or street course. . . . Bourdais set track qualifying records last year with a time of 66.886 seconds and a speed of 105.924 mph. . . . The winner of this race has gone on to claim the series title 10 times.
- Next race: Grand Prix of Houston, April 22.

TODAY'S LINEUP

1. (1) Sebastien Bourdais	1:07.546	2. (5) Will Power	1:07.885
3. (15) Simon Pagenaud	1:07.888	4. (8) Alex Tagliani	1:08.047
5. (2) Graham Rahal	1:08.170	6. (14) Robert Doornbos	1:08.183
7. (9) Justin Wilson	1:08.200	8. (19) Bruno Junqueira	1:08.430
9. (22) Tristan Gommendy	1:08.527	10. (7) Mario Dominguez	1:08.564
11. (21) Neel Jani	1:08.809	12. (42) Matt Halliday	1:08.864
13. (4) Dan Clarke	1:08.890	14. (3) Oriol Servia	1:08.922
15. (28) Ryan Dalziel	1:09.070	16. (29) Alex Figge	1:09.151
17. (11) Katherine Legge	1:09.766	Note-Time in minutes; ()=car number	

Despite falling to No. 2, Australian driver earns a front-row spot for the Grand Prix of Long Beach. Tracy suffers injury and won't race.

By Jim Peltz
Times Staff Writer

His name might sound like a publicist's bad joke, but there's nothing funny about how quickly Will Power is proving he belongs in the big leagues of open-wheel auto racing.

After being named rookie of the year in the Champ Car World Series in 2006, Power, 26, started this season by winning the series' opening race on the streets of Las Vegas a week ago.

It was his first Champ Car victory and the first series win by an Australian.

Power, who drives the No. 5 car for Team Australia, then won the provisional pole for today's 33rd Toyota Grand Prix of Long Beach.

He narrowly lost it in the second round of qualifying Saturday to Sebastien Bourdais, who has won the Long Beach race the last two years and the series title for three consecutive years.

But Power will start second, and he's now had front-row starting berths in three of the last four races, going back to the end of last season.

Bourdais, of the Newman-Haas-Lanigan team, ran a lap at 104.889 mph to win the pole. Power ran 104.858 mph. Power's teammate, French rookie Simon Pagenaud, qualified third.

"We were a little frustrated after yesterday," Bourdais said. "But today was a different day. We made some improvements."

Meanwhile, four-time Long Beach winner Paul Tracy will miss today's race after suffering a broken vertebra in his lower back in a crash during practice Saturday.

The 38-year-old Tracy, who won the series title in 2003, lost control of his No. 3 Forsythe Championship Racing car and slid nose-first into a concrete barrier.

Forsythe chose Spaniard Oriol Servia, a Champ Car driver who was without a full-time ride this year, to replace Tracy for Saturday's qualifying and today's race. He will start 14th in the 17-car field.

Power is leading a youth movement that is challenging

He grew up in Toowoomba in Queensland, Australia, near the Surfers Paradise street course where Champ Car and its predecessor, CART, have raced since

to keep an even keel.

Gore said, "While the team is still on a high after last week's win in Las Vegas, everyone has knuckled down and focused on

Darrell Ingham Getty Images

CONSISTENCY: *Will Power, who had won the provisional pole for the Grand Prix of Long Beach, has been in the front row of three of the last four Champ Car races going back to the end of last season.*

Dave Waters Associated Press

SCARY MOMENT: *Paul Tracy is helped by race safety crew members after crashing in practice and suffering broken vertebra.*

Polesitter Sebastian Bourdais won his third straight, followed by Will Power and Simon Pagenaud. Bourdais led for the first 31 laps, then Servia driving Tracy's backup car led for seven. Tristan Gommendy led for 13 laps until Bourdais took over on lap 68. He led a total of 58 laps and collected $75,000..

There were three cautions; on lap nine when Dominguez crashed, on lap 54 when Pagenaud crashed and on lap 72 when Figge crashed.

Matt Sayles / AP

Sebastien Bourdais rounds the track on his way to winning the Toyota Grand Prix of Long Beach on Sunday. The Frenchman won the race on the downtown streets of Long Beach for the third straight year.

Bourdais cruises to third straight Long Beach win

This weekend was also the third round of the American Le Mans Series; won by Romain Dumas and Timo Bernhard in the Penske Porsche RS with the second Penske Porsche, driven by Sascha Maasen and Ryan Briscoe in second. Third was the Dyson Porsche RS driven by Butch Leitzinger and Andy Wallace.

Dave Mirra won the Toyota Celebrity Race ahead of Martina Navritilova and Joshua Morrow, The first professional, Mike Skinner, was fourth. Hugh Hefner was the grand marshal. Raphael Matos won the Champ Car Atlantic Race with Jonathan Bomarito in

second and Robert Wickens in third. Eric Currran won the Speed GT Race ahead of Andy Pilgrim and Lou Gigliotti. Rhys Millen and Tanner Foust won the EZ Lube Drift Challenge.

(See all the race results on the attached DVD)

Porsche reigns in Le Mans race

By MARTIN HENDERSON
Times Staff Writer

Confusion reigned at the end of the inaugural American Le Mans Series sports car race Saturday at the Toyota Grand Prix of Long Beach, but one thing was clear: The Audi R10 TDI, unbeaten in 10 races since its introduction, including Le Mans, was taken to school by Porsche.

The Audi excelled while it was streaking down Shoreline Drive on the 1.97-mile circuit though the city streets, but the famous Turn 11 hairpin proved troublesome for the series' dominant car in its marquee class, LMP1.

The Audi was credited with a class win, but six cars from the secondary LMP2 class — four Porsches and two Acuras — finished ahead of the previously undefeated Audi co-driven by **Allan McNish** and **Rinaldo Capello**.

Porsche swept the podium, a series first. Owner **Roger Penske's** Porsche RS Spyders finished 1-2 as **Romain Dumas** took the checkered flag first, followed by **Sascha Maassen** a scant 0.76 of a second later. Dumas was teamed with **Timo Ryan Briscoe**.

Dyson Racing's **Butch Leitzinger** took third, 13.4 seconds behind the winner. He was paired with **Andy Wallace**.

"Every run of victories has to come to an end at some point," McNish said. "We have to keep it all in perspective."

The results weren't finalized until well after the race, scheduled for 1 hour 40 minutes. At issue was whether Dumas had spent too much time in the car. Rules stipulate a driver can't be in the car for more than 70% of the race, disqualification being the penalty.

Dumas took over for co-driver Bernhard during a caution period between 30 and 31 minutes into the race, putting the percentage into question, but the result was approved after sorting out pit time and extra time, which do not count toward total driver time, according to ALMS spokesman **Bob Dickinson**.

The race might have gone to Andretti Green Racing drivers **Bryan Herta** and pole-sitter **Dario Franchitti**, but Franchitti did not pit during the race's lone caution flag which meant the green flag conditions. They finished sixth.

Brazilian **Raphael Matos**, who won last week's Atlantic race in Las Vegas, won the pole for today's preliminary to the Grand Prix of Long Beach with a track-record qualifying lap at 93.561 mph.

Dave Mirra, BMX rider and the most dominant athlete in X Games history, was the unsurprising winner of the 31st Toyota Pro-Celebrity race.

The race was stopped after eight of 10 scheduled laps because of a crash involving "Star Wars" director **George Lucas** and charity auction winner **Annamarie Dean**.

Mirra finished ahead of second-place **Martina Navratilova**, the tennis Hall of Famer.

The margin of victory was 7.4 seconds with an average speed of 65.541 mph in identically prepared Toyota Scion tCs. NASCAR truck series driver **Mike Skinner** led the professionals, finishing fourth overall, behind actor **Joshua Morrow**.

BALANCE RESTORED

Read the Autoweek article on the attached DVD and watch the video on our you tube channel - www.youtube.com@racinghistoryproject2607

IMSA Race

Formula Atlantic Race

ANNE CUSACK *Los Angeles Tim*

NOT ENOUGH: *Will Power, foreground, stays ahead of his teammate Simon Pagenaud, background, who later spun, bringing out a caution that closed Sebastien Bourdais' lead.*

Katherine Legge

Justin Wilson

Formula Drift

Toyota Celebrity Race

Sebastian Bourdais

Oriol Servia

Will Power

Simon Pagenaud

Mario Dominguez

Graham Rahal

Alex Tagliani

Paul Tracy

Bourdais, Power and Pagenaud

2008

The 2008 Toyota Grand Prix of Long Beach, round three of the "Indy Car Series", was held on Aprll 20, 2008 in front of a crowd of over 100,000 with over 200,000 for the weekend. The race conflicted with the Indy Japan 300 due to the merger of the IRL and Champ Car series' and only 20 cars took part. This would be the last Champ Car event as the Grand Prix Association of Long Beach signed a long term agreement with Indy Car, beginning in 2009.

This years Meet The Press event was held at the Museum of Latin American Art. The Long Beach Motorsports Walk of Fame, along Pine Avenue in downtown, inducted three members in 2008; Mario Andretti, Gary Gabelich and Parnelli Jones.

The Automobile Club Lifestyle Expo featured motorcycle stunts, driving simulators, alternative energy vehicles. and vendors of performance equipment.

Friday was practice and qualifying for all plus the Tecate Light Fiesta with "Aleks Syntek" and the Tecate Miss Toyota Grand Prix of Long Beach Competition, A busy Saturday featured the IMSA Tequila Patron Grand Am and the Toyota Celebrity Race at 11:30 AM, Formula Drift and drifting demonstrations and the Rock-N-Roar concert with "Pennywise" at 6:00 PM. Over 50 police were needed to quell the brief riot at the Long Beach Convention Center after crowds were turned away from the concert.

Sunday, race day, had the Champ Car Mazda Atlantic Race at 10:35 AM, the CART Race at 1:00 PM. Formula Drift at 3:20 PM and the Speed World Challenge Race at 4:00 PM. Bobby Rahal was the grand marshal. The race was televised on ESPN 2 with reporters Scott Goodyear, Danica Patrick, Jack Arute and Jamie Little.

Wilson wins last Champ pole

BY MIKE HARRIS
ASSOCIATED PRESS

LONG BEACH, — Justin Wilson isn't just competing with the rest of the field at the Toyota Grand Prix of Long Beach, he's racing against a driver who isn't even here.

When Sebastien Bourdais left the Champ Car World Series at the end of the 2007 season to race

This was the first time a standing start was used. Justin Wilson qualified on the pole but Will Power, who started third, won the race, collecting $97,500. Second was rookie Franck Montagny, five seconds back, with Mario Dominguez in third. Power took the lead at the start while Servia stalled at the starting line, bringing out a yellow. Wilson ran second until something broke on lap 13. He finished 19th. Mario Moraes hit the turn two wall bringing out the next yellow while Power continued to lead with Wilson second until he dropped out. Nelson Phillipe created the third yellow when his car quit in turn 11. Power led 81 out of 83 laps.

SPENCER WEINER *Los Angeles Times*

GRAND FINALE: *Will Power wins at Long Beach, the last Champ Car World Series event.*

Power cruises at Long Beach

[*Grand Prix, from Page D1*]
major U.S.-sanctioned open-wheel race.

She won in her 50th start with a fuel strategy that enabled her Andretti Green Racing car to stay on the track at the finish while other leaders had to pit for fuel.

Then, in Long Beach, Power dominated the 20-car field and the 27-year-old Australian closed Champ Car's books with his third series win.

Rookie Franck Montagny was second, five seconds be-

Race results

Lap length: 1.968 miles (start position in parentheses):

PL	ST	DRIVER	LAPS
1.	(4)	Will Power	83*
2.	(6)	Franck Montagny	83
3.	(10)	Mario Dominguez	83
4.	(8)	Enrique Bernoldi	83
5.	(12)	Oriol Servia	83
6.	(3)	Franck Perera	83
7.	(2)	Alex Tagliani	83

the 13th lap when something on his Newman-Haas-Lanigan car broke as he drove down Shoreline Drive, ending his race. He finished 19th.

And Rahal, the 19-year-old son of former Indy 500 winner Bobby Rahal, was running in the top 10 until the final lap, when he spun out while trying to end with a flourish. He finished 13th.

Rahal, whose father was the race's grand marshal, dropped to ninth in IndyCar points. He made history as well two weeks

Simone de Silvestro won the Atlantic Challenge Race with Alan Sciato in second and Kevin Lacroix in third. Lucas Luhr and Marco Werner in an Audi R10won the 71 lap IMSA American LeMans Race with teammates Frank Biela and Emanuele Pirro second and the Courage Acura of David Brabkham and Scott Sharp third. The Toyota Celebrity Race was won by Jamie Little who was also the first amateur, with Mike Skinner. The first professional, second and Daniel Goddard third. Brandon Davis won the Speed World Challenge Race.

(See all the race results on the attached DVD)

COMPETITION

DARKNESS FALLS

Read the Autoweek article on the attached DVD and watch the video
on our you tube channel - www.youtube.com@racinghistoryproject2607

'I LIKE TO GO BIG': *Vaughn Gittin Jr. is ready to drift at Irwindale, this weekend's D1 Grand Prix host.*

[*Drifting, from Page 29*]
cars in the process. "It's the same story, except instead of karate, it's cars."

If the film signals drifting's breach into the American mainstream, in Japan it has been an organized sport for about a decade and an underground subculture for decades before that. Indeed, to hear enthusiasts tell it, drifting's origins are shrouded in mystery worthy of Zatoichi the Blind Swordsman.

It apparently evolved out of illegal road racing on winding mountain roads (or *touge*) in the 1960s — nobody knows where or

market sport-compact car craze.

In a trajectory similar to hot rodding's morphing into sanctioned drag racing in the United States in the 1950s, drifting went legit. In 2001, magazine publisher Daijiro Inada and pro driver Keiichi Tsuchiya founded the D1 Grand Prix series, which you might think of as the Formula 1 of going sideways.

And then it jumped the pond.

Southern California was already heavily invested in noisy Japanese imports, as any number of frantic suburbanites could have told you. The area is ground zero for the sport-compact/tuner

Drifting events

D1 Professional Drift Grand Prix Series

Where: Irwindale Speedway, 500 Speedway Drive, Irwindale

When: Qualifying noon to 6 p.m. Friday; gates open at 9 a.m. Saturday, practice session at 10 a.m., opening ceremony at 2:30 p.m., final competition at 6:45 p.m.

Price: Friday, free; Saturday, $25 in advance, $30 at the door

Info: www.d1gp.com

Formula Drift Championship Series

Where: Long Beach Grand Prix course in downtown Long Beach

When: 10 a.m. to 4 p.m. April 2

Price: $20

Info: www.formuladrift.com

Also

What: Drifting events during Long Beach Grand Prix weekend

When: 5 to 5:25 p.m. April 7, 3:10 to 3:30 p.m. April 8, 3:20 to 3:40 p.m. April 9

Price: $55 for three-day general admission to all Long Beach Grand Prix events

Info: www.longbeachgp.com

More events on the Web

www.driftday.com, www.driftcircuit.org, www.justdrift.com, www.driftbuffet.com

City of Long Beach City Manager Pat West presenting Long Beach Motorsports Walk of Fame plaque to Mario Andretti on April 17, 2008

Gears will shift at Long Beach

Sunday's race will pay tribute to Champ Car series in its last sanctioned event as it fully merges under the IRL's IndyCar banner.

By Jim Peltz
Times Staff Writer

Both by design and happenstance, the 34th Toyota Grand Prix of Long Beach on Sunday will symbolize the past, present and future of American open-wheel auto racing.

The race through the city's downtown streets will be the last sanctioned by the Champ Car World Series, which made years included such legendary drivers as Mario Andretti and Al Unser Jr., along with four-time winner Paul Tracy and Sebastien Bourdais, who won the last three races before moving this season to Formula One.

But this year's event isn't just about nostalgia. It's also a key race in the championship ambitions of those Champ Car drivers who have migrated to the IRL, because they'll still earn IndyCar Series points based on their Long Beach finishes.

They include 19-year-old Graham Rahal, who stunned the sport two weeks ago by becoming the youngest winner in U.S. open-wheel racing history with a victory on another street course, in St. Petersburg, Fla. Rahal, the son of 1986 India-

SUNDAY'S RACE
Grand Prix of Long Beach
1 p.m. (ESPN2, delayed, 2:30).

IndyCar team will race in Japan on Saturday (tonight, Pacific time), and then he will fly back to Long Beach.

The Long Beach circuit is on the 1.97-mile, 11-turn course on the city's seaside streets that includes a long stretch of Shoreline Drive.

The Rahals represent the future of open-wheel racing. After Sunday, they will be racing in the same series, which will return to Long Beach next year with other IndyCar drivers such as Dan Wheldon, Helio Castroneves, Tony Kanaan, Danica Patrick and Marco An-

time; it's nice to be racing, but with regret," Tagliani said.

Another driver is Jimmy Vasser, who won Long Beach (and the series title) in 1996 before retiring to become a team owner. He's coming out of retirement to drive this race for KV Racing Technology, the team he co-owns.

Long Beach uses a two-day qualifying format for the grand prix. The driver with the fastest lap today is at least assured a spot on the two-car front row, and Saturday's second round determines the pole-sitter. The weekend also includes several support series that race on the same course, and all start practicing and/or qualifying today.

Saturday features a race between celebrities and racers driving identical Toyota Sci-

Associated Press

REMEMBER KIDS, DON'T DRINK AND DRIVE: ESPN racing reporter Jamie Little, *left, takes a celebratory swig of champagne after winning the overall title at the Pro/Celebrity Race at the Long Beach Grand Prix on Saturday. NASCAR Sprint Cup driver Mike Skinner, right, won the Pro category.*

Joe Benson was broadcasting from the Bubba Gump Shrimp Company

David Brabham leads Frank Biela in the IMSA Race

Jimmy Vasser

Justin Wilson

Winners and Champagne

Paul Tracy

Alex Tagliani

Franck Montagny

Antonio Pizzonia

Roberto Moreno

Enrique Bernoldi

Michael Lewis / Tomy Drissi in the Judd

Celebrate 25 years
of Champ Car racing!
Toyota Grand Prix
of Long Beach
April 18-20
HOT GREEN! Revisit Long Beach's Champ Car past and enjoy 6 racing series including the final Champ Car Race, Toyota Pro/Celebrity, EZ Lube Team Drift, Tequila Patrón American Le Mans Series and Imperial Capital Bank Atlantics.
COOL GREEN! Tecate Light Fiesta Friday Concert ("Aleks Syntek"), Saturday Rock-N-Roar Concert ("Pennywise") and Tecate Miss Toyota Grand Prix of Long Beach Pageant....FREE with race ticket!
GREEN HOUSE! The expanded Lifestyle Expo will showcase renewable energy sources & alternative energy vehicles.
PLUS...all the extras that make the Toyota Grand Prix of Long Beach America's 'Spring Brake!'
Tickets/Information:
(888) 82-SPEED
www.GPLB.com
ticketmaster
TOYOTA GRAND PRIX
OF LONG BEACH
APRIL 18-20, 2008

2009

The 2009 Toyota Grand Prix of Long Beach, round two of the Indy Car Series, the result of the merger with Champ Car, was held on Aprll 19, 2009 in front of a crowd of over 100,000.

Friday was practice and qualifying for all plus the Automobile Club Lifestyle Expo featured motorcycle stunts, driving simulators, alternative energy vehicles and vendors of performance equipment. A busy Saturday featured the IMSA Tequila Patron Grand Am Race, then the Toyota Celebrity Race at 11:30 AM. Later came Formula Drift and drifting demonstrations and the Rock-N-Roar concert at 6:00 PM. Sunday had qualifying, practice, team drifting, the Firestone Indy Lights Race at 9:40 AM, then various pre race activities with the Indy Car Race at 1:20 PM. Following that was the Speed World Challenge Race at 4:15 PM.

The race was televised on Versus, which used to be the Outdoor Life Channel, with Jack Arute, Robbie Buhl and John Beekhuis as reporters. Bob Jenkins was the lead announcer. Al Unser Jr. was grand marshal.

Power takes Long Beach pole

By Mike Harris
AP auto racing writer

LONG BEACH — Nothing seems to slow Will Power down at Long Beach.

Thanks to the return of Helio Castroneves, the Australian driver had to climb out of one of the fastest cars entered in the Toyota Grand Prix of Long Beach and jump into a new car that had not turned a wheel on the track.

No matter. Power drove his new Team Penske No. 12 Dallara to the pole position for today's race.

"I got the (No.) 3 car in pretty

Dario Franchitti won on a hot day with pole qualifier Will Power second, without a working radio, three seconds back, and Tony Kanaan in third. Franchitti had a bad start, dropping to fourth while Power led. A multi car crash that started when Dixon hit Rahal involved Wilson, Dixon, Rahal, Manning Moraes and Mutoh. Mutoh, Manning, Dixon and Rahal were

able to continue. Viso and Conway had their own accidents, generating yellows. Helio Castroneves had just been acquitted on a tax evasion charge and got his ride back with Penske, relegating Power to a spare car.

Dave Waters / AP

Dario Franchitti crosses the finish line to win the IndyCar Series Grand Prix of Long Beach on Sunday.

Franchitti's return to IndyCar pays off with Long Beach win

Driver gives his former NASCAR boss Ganassi

NASCAR ride, Ganassi decided to team Franchitti with 2008 Indy winner and series champion Scott Dixon and the move has paid off this season with a

The Toyota Celebrity Race was won by Al Unser Jr., first of the professionals, followed by Johnny Benson Jr and in third, the first amateur, Toyota dealer Tom Rudnai. Simon Pagenaud and Gil de Ferran won the Tequila Patron American Le Mans Race in an Acura followed by teammates Scott Sharp and David Brabham. Luis Diaz and Adrian Fernandez were third in a Courage / Acura. The Indy Lights Race was won by J.R. Hildebrand with Richard Phillipe in second and James Hinchcliffe third.

(See all the race results on the attached DVD)

Castroneves doesn't show any rust

A day after being acquitted of charges of tax evasion, the IndyCar star qualifies eighth at Long Beach.

JIM PELTZ

In between well-wishers' hugs and handshakes on pit road, Helio Castroneves once again strapped himself into his No. 3 red-and-white Team Penske race car Saturday.

Castroneves' ready grin also was back, but knowing his driver was feeling butterflies ahead of practicing for today's Toyota Grand Prix of Long Beach, Team Penske President Tim Cindric figured a touch of humor couldn't hurt. "There's a lot of people watching," Cindric told his Brazilian driver, "so don't stall it."

They watched because only one day earlier, a jury acquitted Castroneves of federal tax-evasion charges that threatened to end the career of one of the IndyCar Series' most successful and popular drivers.

Instead, Castroneves — who won the Long Beach race in 2001 — practiced and then qualified eighth in the 23-car field despite not having driven the car for several months.

At the end of Castroneves' run, however, he spun and hit the retaining wall in Turn 1. Castroneves complained of a

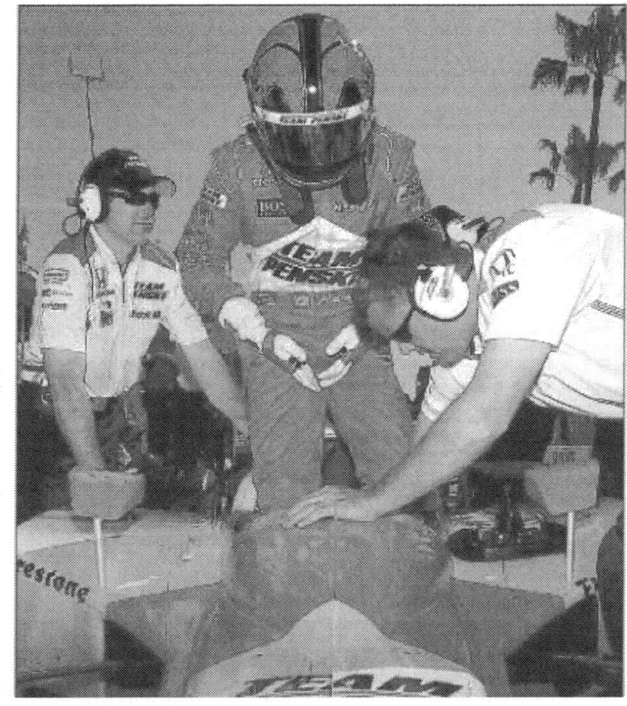

Long Beach Grand Prix

TODAY'S SCHEDULE

Gates open	7 a.m.
Lifestyle Expo opens	8 a.m.
SPEED GT Challenge qualifying	8–8:30 a.m.
IndyCar practice	8:40–9:10 a.m.
Firestone Indy Lights	9:40–10:50 a.m.
Team drifting	11:05–11:30 a.m.
IndyCar pre-race activities	12:30 p.m.
35th Annual Toyota Grand Prix of Long Beach	1:20–3:30 p.m.
SPEED GT Challenge	4:15–5:15 p.m.
Lifestyle Expo closes	5:30 p.m.

THE LINEUP

on streets of Long Beach; lap length: 1.968 miles

PP	No.	Driver	Car	Qualifying	Fastest
1.	12	Will Power	Dallara-Honda	1:09.7107	101.631
2.	10	Dario Franchitti	Dallara-Honda	1:09.8675	101.403
3.	2	Raphael Matos	Dallara-Honda	1:10.2043	100.917
4.	13	E.J. Viso	Dallara-Honda	1:10.2232	100.890
5.	18	Justin Wilson	Dallara-Honda	1:10.2680	100.825
6.	9	Scott Dixon	Dallara-Honda	1:10.4038	100.631
7.	02	Graham Rahal	Dallara-Honda	1:10.0283	101.171
8.	3	Helio Castroneves	Dallara-Honda	1:10.1139	101.047
9.	34	Alex Tagliani	Dallara-Honda	1:10.1278	101.027
10.	6	Ryan Briscoe	Dallara-Honda	1:10.2126	100.905
11.	11	Tony Kanaan	Dallara-Honda	1:10.3032	100.775
12.	21	Ryan Hunter-Reay	Dallara-Honda	1:10.4339	100.588
13.	5	Mario Moraes	Dallara-Honda	1:10.8459	100.003
14.	4	Dan Wheldon	Dallara-Honda	1:10.5503	100.422
15.	06	Robert Doornbos	Dallara-Honda	1:10.9577	99.845
16.	24	Mike Conway	Dallara-Honda	1:10.6059	100.343
17.	27	Hideki Mutoh	Dallara-Honda	1:11.1177	99.621
18.	23	Darren Manning	Dallara-Honda	1:11.2411	99.448

Helio Castroneves

Ryan Briscoe

Ed Carpenter

Darren Manning

Kimball turns to a different road

Rio Mesa High graduate, who has spent much of his racing career in Europe, will drive in Indy Lights car in Long Beach Grand Prix

Michael L. Levitt / LAT Photographic

Charlie Kimball, who grew up in Camarillo, will race in an Indy Lights car at the Long Beach Grand Prix on Sunday.

By Tim Haddock
Correspondent

LONG BEACH — Firestone Indy Lights driver Charlie Kimball doesn't quite own the streets of Long Beach yet, but his family has its name on a road in Ventura.

Kimball is racing in the Indy Lights portion of the Toyota Grand Prix of Long Beach this weekend. It is the first time since he was racing go-karts as a kid that Kimball has been able to race in Southern California. Kimball said he is excited about getting a chance to race in front of his hometown friends and family.

Kimball, 23, grew up in Camarillo and graduated from Rio Mesa High. Kimball Road, off Highway 126 in Ventura, is named after his grandfather and great uncle, who used to own a ranch along that stretch of farmland. His parents own an avocado farm between Fillmore and Santa Paula.

By Kimball's estimation, at least four generations of Kimballs have been growing oranges and avocados in Ventura County. His family roots are strong in that agricultural community and it's a heritage Kimball is proud of.

But the lure of racing cars had a

See KIMBALL on C2

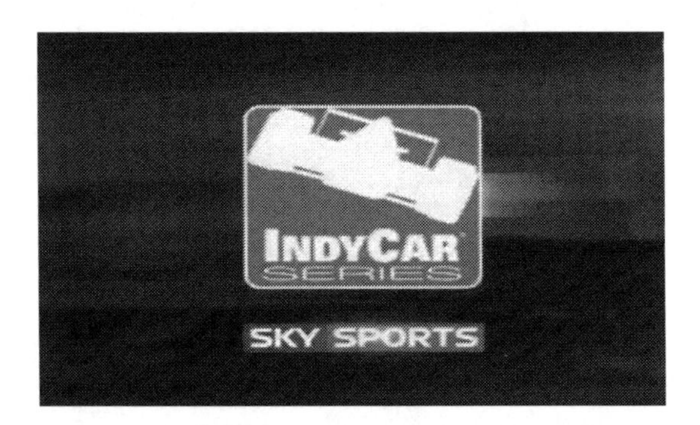

Read the Autoweek article on the attached DVD and watch the video on our you tube channel - www.youtube.com@racinghistoryproject2607

American Le Mans Race

Formula Drift

Toyota Celebrity Race

Kanaan Pouring Champagne on Franchitti

2010

The 2010 Toyota Grand Prix of Long Beach, race four of the Izod Indy Car Series, was held on Aprll 18, 2010 in front of a crowd of 170,000 over the three day event. This was the 25th anniversary of Toyota's sponsorship of the race.

Thursday was Tecate Thursday Thunder on Pine Avenue plus the Miss Toyota Grand Prix Contest,

Friday's concert on the Rock-N-Roar Stage outside the Convention Center featured Tecate Light Fiesta Friday with "Molotov" and "Maldita Vecindad". Saturday featured "Cheap Trick". More musical acts appeared on the stage in the Long Beach Arena all weekend. Practice and qualifying went on all day Friday at the track.

The Automobile Club Lifestyle Expo featured 130 exhibitors, live entertainment, the "Kids Zone", BMX bike action plus the Green Power Preview of alternative vehicles.

Michael Andretti and Danny Sullivan were the 2010 inductees into the Long Beach Motorsports Walk of Fame.

Saturday featured the Toyota Celebrity Race, Formula Drift and the Tequila Patron IMSA American Le Mans Series Race. Sunday, race day, had the Firestone Indy Lights Race at 11:25 AM then the popular driver introduction laps in Toyota Tundras, the Indy Car Race at 1:00 PM followed by the SCCA World Challenge Race.

The race was live on VERSUS with commentators Jack Arute, Robbie Buhl and John Beekhuis. Bob Jenkins was the announcer. The broadcast averaged 400,000 viewers.

Power speeds to win his third straight IRL pole

Associated Press

Will Power earned his third consecutive Indy Racing League pole, giving him a shot at a third win in four races.

The winner in Brazil

Polesitter Will Power got the lead at the start followed by Ryan Hunter Reay, second fastest qualifier, and Justin Wilson. Power made the mistake of hitting the pit speed limiter on lap 17, allowing Hunter Reay and Wilson to get by. On lap 63, Wilson, while lapping Alex Lloyd, bumped him and damaged his front spoiler.

dropping him to third. A caution period on lap 59 when Romancini and Rahal got together bunched up the field but Hunter Reay got away on the restart and won by five seconds over Wilson with Power in third.

Hunter-Reay dominates Grand Prix of Long Beach

Driver wins by 5.6 seconds over second-place Wilson. Power finishes third.

JIM PELTZ

Ryan Hunter-Reay is known as a pretty face of IndyCar racing — in addition to Danica Patrick, of course — in good part because he's an advertising centerpiece of the sport's series sponsor, Izod clothing.

On the racetrack, though, the 29-year-old Hunter-Reay has endured a career that hasn't always been so attractive. He has bounced between different teams and, before Sunday, had won only one race in the last four years.

[See Grand Prix, C6]

Results

(Start position in parentheses)

1. Ryan Hunter-Reay (2)
2. Justin Wilson (3)
3. Will Power (1)
4. Scott Dixon (8)
5. Tony Kanaan (6)

>>>Complete results, C10

GINA FERAZZI Los Angeles Times

TOP THREE: Race winner Ryan Hunter-Reay, right, is joined by second-place finisher Justin Wilson, left, and Will Power, who finished third.

Jimmy Vasser won the Toyota Celebrity Race and the professional class with Tanner Foust in second. The first amateur, Brian Austin Green, was third. The World Challenge Race was won by Kuno Wittmer with Dino Crescentini in second and Boris Said in third. Simon Pagenaud and David Brabham won the American Le Mans Series Race in a Courage / Acura with Adrian Fernandez and Harold Primat in a Lola / Aston Martin in second. Klaus Grai and Greg Pickett were third in a Porsche RS. James Hinchcliffe won in Indy Lights with Charlie Kimball second and Jean Karl Vernay in third.

(See all the race results on the attached DVD)

Ryan Hunter Reay and the Team

World Challenge

American Le Mans Race

HUNTER-REAY WINS LONG BEACH

Read the Autoweek article on the attached DVD and watch the video on our youtube channel - www.youtube.com@racinghistoryproject2607

Toyota Celebrity Race

Scott Dixon

Justin Wilson

Tony Kanaan

Helio Castroneves

Dario Franchitti

Milka Duno

Marco Andretti

Danica Patrick

Ryan Hunter Ray. Justin Wilson and Will Power

Miss Toyota Grand Prix and Runner Ups

GINA FERAZZI Los Angeles Times

FUELING TOURISM: The Long Beach Grand Prix gives the city a chance to show itself off on national TV.

For Long Beach, race is a 'marketing megaphone'

The Grand Prix attracts hordes of travelers and a national spotlight

DAVID SARNO

Two dozen racing machines zoomed past the grandstands along Long Beach's Shoreline Drive on Sunday, showcasing more than enough horsepower to tow the nearby 91,000-ton Queen Mary out of its mooring.

And in drawing about downtown area for conferences and conventions, according to a study last year by the economics department at Cal State Long Beach.

But this weekend the city's engine was purring. Downtown hotels sold out for days in a row, and local bars and restaurants said they were having their best stretch of 2010.

the city lost 50,000 aerospace and defense jobs and skidded to what Foster called "depression-level conditions."

In the years since, the city has spent heavily to redefine itself as a convention-friendly destination, one of the few large California cities with a downtown along the water. Long Beach hosts dozens of sports and special events ev-

As cars roared by outside, local rock bands played on an elevated stage in the convention center a few feet from where children were bouncing on an inflatable play area. From rows of booths, vendors sold diamond rings and panoramic photographs of the downtown marina. And Toyota Motor Corp. — the event's main sponsor — showed off a

TOYOTA GRAND PRIX

Three days of partying and professional racing take over the streets of Long Beach next weekend

The 36th annual Toyota Grand Prix of Long Beach roars into action next weekend, Friday, April 16 through Sunday, April 18, featuring the fourth round of the 2010 IZOD IndyCar Series. (There will also be a few earlier events.)

The race takes place on the streets of downtown Long Beach surrounding the Long Beach Convention and Entertainment Center and encompassing the Aquarium of the Pacific and a portion of the Pike at Rainbow Harbor complex. The start/finish line is on Shoreline Drive.

The weekend event draws about 175,000 spectators each year, making it one of the largest paid spectator special events in Southern California.

Featured race: The featured race is Sunday afternoon's IZOD IndyCar Series event, which includes an international field of world-class drivers such as: Indy 500 winner Helio Castroneves, 2009 Long Beach winner and Series champion Dario Franchitti, Danica Patrick, Marco Andretti and Scott Dixon. Five other racing events will also be run.

Toyota Pro/Celebrity Race: The 34th edition of this fan favorite race features stars from the sports and entertainment worlds battling professional drivers in identically prepared Scion tCs. Keanu Reeves returns to defend his 2009 win along with 18 other celebrities and pros. The race will take place midday on Saturday, April 17.

Tequila Patrón American Le Mans Series at Long Beach race: This race features the most technically advanced sports cars in competition. Practice and qualifying for ALMS will be held on Friday, April 16, and the race will

2011

The 2011 Toyota Grand Prix of Long Beach, race three of the Izod Indy Car Series, was held on Aprll 17, 2011 in front of a crowd of 170,000 over the three day event.

Thursday was the Long Beach Walk of Fame Ceremony at 11:00 AM. Inducted Cip Ganassi Racing and Jimmy Vasser. The Miss Toyota Grand Prix Contest took place at 6:00 PM

Friday was practice and qualifying, an autograph session with Indy Car drivers at 4:00 PM and a Tecate Light Concert with "Moderatto" and "Fobia" that evening.

Saturday had another Tecate Light Concert with "John Kay and Steppenwolf" presented by KLOS, go kart rides, the Mothers Exotic Car Paddock and the Lifestyle Expo with a rock climbing wall, kids electric car rides, BMX and skateboard demonstrations and an indoor zip line. Racing included the Toyota Pro Celebrity Race at 11:40 AM and the American Le Mans Series Race at 4:30 PM. The evening had a concert with "John Kay and Steppewolf" at 6:45 PM.

On Sunday, the Indy Car Race started at 1:15 PM followed by team drifting at 3:40 PM.

The race was broadcast on Versus with reporters Wally Dallenbach Jr, Jon Beekhuis and Robin Miller. Bob Jenkins was the announcer.

GRAND PRIX NOTES

Power wins pole at Long Beach

JIM PELTZ

Will Power won the pole position for the Toyota Grand Prix of Long Beach, the third consecutive year that the Australian has started first in the Izod IndyCar Series race.

Power also has captured the pole for all three of the series' races so far this season, and he won the second race, last weekend, at Barber Motorsports Park in Birmingham, Ala.

The Team Penske driver Prix last year, was second with a lap of 102.469 mph.

"We won from there [in second] last year so hopefully we can repeat," Hunter-Reay said.

Mike Conway qualified third in the 27-car field, and **Oriol Servia** was fourth. Power's Penske teammates **Helio Castroneves** and **Ryan Briscoe** finished sixth and 12th, respectively. **Danica Patrick** qualified 20th. **Charlie Kimball** of Camarillo was 24th.

Power won the Long Beach Grand Prix in 2008, reigning IndyCar champion, was seventh, and his teammate **Scott Dixon** was eighth.

Franchitti also won this year's season opener in St. Petersburg, Fla.

"Everyone is surprised" the Ganassi team didn't qualify higher, said **Justin Wilson**, who will start fifth. "They're always so strong. It just shows you, you can't miss a beat."

Wheldon entry

Dan Wheldon, who won the Indianapolis 500 and the

Will Power qualified on the pole for the third consecutive year but ended up tenth. He led on the faster "red compound" tires until a spin, first by Wilson, then by Simona de Silvestro brought out a caution. A later incident with Castroneves put an end to Power's shot at winning. Mike Conway won, six seconds ahead of Ryan Briscoe with Dario Franchitti third. 17 cars finished on the lead lap. Briscoe led 35 laps while Power led 29. There were only three caution periods

Conway wins Long Beach Grand Prix

Long Beach Grand Prix

The top eight at the Indy Racing League event through the streets of Long Beach (starting position in parentheses):

1.	Mike Conway (3)
2.	Ryan Briscoe (12)
3.	Dario Franchitti (7)
4.	James Hinchcliffe (11)

The Englishman passes Briscoe with 14 laps left to capture his first IndyCar victory.

JIM PELTZ

Few IndyCar drivers are more reserved and soft-spoken than Mike Conway, a 27-

rific crash at last year's Indianapolis 500, won the Toyota Grand Prix of Long Beach — his first victory in the Izod IndyCar Series in 26 starts.

"The car was just on fire, really, at the end," said Conway, who drives for Andretti Autosport, the team co-owned by former racer Michael Andretti. "Before I knew it I was in the lead."

Conway passed Ryan

"Kudos to [Conway]; it's a pretty special place for him to get his first [win]," Briscoe said of the famed 11-turn, 1.97-mile Long Beach seaside circuit. "Conway was lightning fast; I really had nothing for him."

Reigning IndyCar champion Dario Franchitti finished third and rookie James Hinchcliffe was fourth. Danica Patrick, a

Klaus Graf and Lucas Luhr won the IMSA American Le Mans Series Race in a Lola / Aston Martin. Second was Guy Smith and Chris Dyson in a Lola / Mazda followed by Ricardo Gonzalez and Gunnar Jeanette in an Oreca / Chevrolet. William Fichtner won the Toyota Celebrity Race followed by Ken Gushi and Michael Trucco. Brandon Davis won the World Challenge Race followed by James Sofronas and Jason Daskalos. Conor Daly won the Indy Lights Race followed by Esteban Guerrieri and Stefan Wilson.

(See all the race results on the attached DVD)

Read the Autoweek article on the attached DVD and watch the video on our youtube channel - www.youtube.com@racinghistoryproject2607

Formula Drift

Ryan Briscoe

Helio Castroneves

Ryan Hunter Reay

Sebastian Bourdais

Mike Conway

Tony Kanaan

Simona de Silvestro

Takuma Sato

American Le Mans Series

Toyota Celebrity Race

World Challenge

Conway, Briscoe and Franchitti

2012

The 2012 Toyota Grand Prix of Long Beach, race three of the Izod Indy Car Series, was held on Aprll 15, 2012 in front of a crowd of 170,000 over the three day event.

Thursday was the Long Beach Walk of Fame Ceremony at 11:00 AM. Inducted were Galles Racing, Scott Pruett and Adrian Fernandez. Then at 6:00 PM there was the Miss Toyota Grand Prix Contest,

Friday, although it rained, was practice and qualifying, an autograph session with Indy Car drivers at 4:00 PM and a Tecate Light Fiesta Friday Concert with "Belanova" that evening.

Saturday had the Rock-N-Roar Concert with "Joan Jett and the Blackhearts" plus the family fun zone and the Lifestyle Expo with a rock climbing wall, kids electric car rides, BMX and skateboard demonstrations and the Green Power Prix. Racing included the Toyota Pro Celebrity Race at 11:40 AM, with the national anthem sung by LeAnn Rimes and the Tequila Patron IMSA American Le Mans Series race at 4:30 PM.

Sunday, race day, the Star Spangled Banner was sung by Taylor Dane and there was a C-17 Flyover. The Indy Lights race started at 10:45 AM, the Indy Car Race 1:15 PM and the Pirelli World Challenge Race at 4:15 PM..

The race was broadcast on NBC Sports with reporters Wally Dallenbach Jr., Townsend Bell, Jon Beekhuis and Robin Miller. Bob Jenkins was the announcer.

Briscoe wins Long Beach pole, will start in No. 11 spot

By Jenna Fryer
Associated Press

LONG BEACH — Ryan Briscoe kept Penske Racing perfect so far this season by winning the pole for today's Toyota Grand Prix of Long Beach.

He won't be there for very long.

A decision by Chevrolet

cided it was in the best interest of all our Chevy teams, and the Long Beach Grand Prix to swap the engines out beforehand, knowing the grid penalty will be difficult to overcome in the race."

A total of 14 drivers were pushed back in the field for engine changes — three Lotus drivers also made

Pole winner will drop to 11th for race's start

Engine changes result in penalties for 14 drivers

LONG BEACH, Calif. (AP) — Ryan Briscoe kept Penske Racing perfect so far this season by winning the pole for the Toyota Grand Prix of Long Beach.

He won't be there for very long.

A decision by Chevrolet to yank the engines from all 11 of its teams because of concerns the engines wouldn't last throughout the race meant all the Chevy drivers knew they

Ryan Briscoe celebrates after qualifying Saturday at the Toyota Grand Prix of Long Beach in Long Beach, Calif. Briscoe took the pole position. MARK J. TERRILL / ASSOCIATED PRESS

Before the first practice, all Chevrolet entries were penalized because of unauthorized engine changes. This penalty moved the polesitter, Ryan Briscoe, from starting in the first position down to the 11th spot. Dario Franchitti, in the quickest non Chevrolet car, started in the first position. Still, Will Power won with Simon Pagenaud second, only .86 seconds back and James Hinchcliffe third. On a two stop race strategy, Power was able to stretch his fuel to run the last 31 laps. Takuma Sato ran in the top three most of the day until an incident with Hunter Reay on the last lap. Justin Wilson led for 15 laps but was forced to fuel in the final laps. Franchitti led the first four laps but after contact with Newgarden which put him out, he struggled and dropped out three laps from the end. Marco Andretti and Graham Rahal crashed and during thst yellow flag Dixon's car quit.

Adam Carolla won the Toyota Celebrity race with Hill Harper second and Biff Gordon third. Klaus Graf and Lucas Luhr won the IMSA American Le Mans series race in an HPD / Honda with Chris Dyson and Guy Smith second in a Lola / Mazda and Ryan Dalziel and Alexpopov third in an Oreca / Chevrolet. Esteban Guerrieri won the Indy Lights Race with Sebastian Saavedra second and Tristan Vautier third. The Pirelli World Challenge Race GT Class was won by Andy Pilgrim with Randy Pobst second and Johnny O'Connell third. The GTS class was won by Jack Baldwin with Justin Bell second and Colin Braun third..

(See all the race results on the attached DVD)

Read the Long Beach and Indy Car article on the attached DVD and watch the video on our youtube channel - www.youtube.com@racinghistoryproject2607

Miss Long Beach and the Tecate Light Girls

E.J. Viso

Charlie Kimball

Helio Castroneves

Ed Carpenter

Ryan Briscoe

James Jakes

Justin Wilson

Simon Pagenaud

Associated Press

Will Power, in the Verizon car, races in front of the pack, as the Long Beach convention center stands in the background, during the IndyCar Series' Grand Prix of Long Beach auto race, Sunday in Long Beach. Power won the race.

Power, Pagenaud and Hinchcliffe

Toyota Celebrity Race

IMSA American Le Mans Race

TURBO CHARGED excitement!
APRIL 13-15
38TH ANNUAL TOYOTA
GRAND PRIX
of Long Beach
• New turbocharged engines and chassis of the IZOD IndyCar Series with stars like Dario Franchitti, Helio Castroneves and Will Power
• Five more races including the Toyota Pro/Celebrity Race, Tequila Patrón American Le Mans Series and drifting
• Lifestyle Expo, Green Power Prix-View and Family Fun Zone
• Two FREE Tecate Light concerts:
- Fiesta Friday with Belanova
- Saturday Rock-N-Roar starring rock legend Joan Jett and the Blackhearts
LOTUS
FOR MORE INFO OR TO PURCHASE TICKETS ONLINE
WWW.GPLB.COM
(888) 82-SPEED (77333)
IZOD
Joan Jett and the Blackhearts
Follow us:

2013

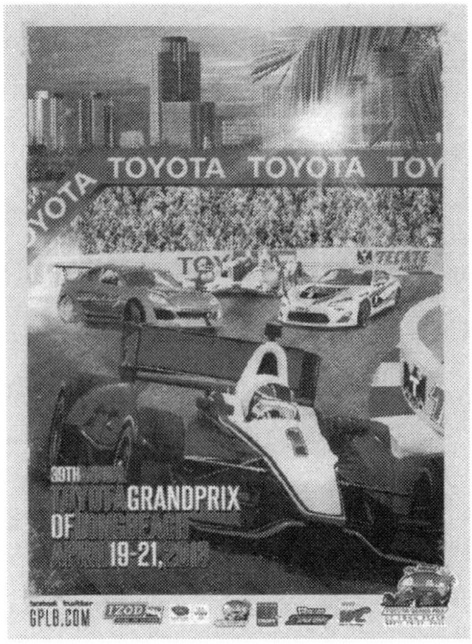

The 2013 Toyota Grand Prix of Long Beach, race three of the Indy Car season, was held on Aprll 21, 2013.

The Lifestyle Expo featured the “”Family Fun Zone” with rock climbing walls, racing simulators and kids three to six driving little racecars. Older kids and adults can race go karts on top of the Long Beach Arena Parking Garage. The Food Truck Experience was located behind Grandstand 31, south of the front straightaway.

Friday was practice and qualifying for all plus ihe Indy Car Autograph Session in the paddock. At 6:45 PM, the Tecate Light Fiesta Friday Concert featured “Moderatto” and “Moenia”.

Saturday featured the Toyota Celebrity Race at 11:40 AM, the Tequila Patron American Le Mans Race at 4:30 and from 7:00 PM to 9:00 PM, the Motegi Super Drift Challenge. The Rock ‘n Roar Concert with Bret Michaels took place at 6:45 PM..

Sunday, race day, had the Mothers Exotic Car Paddock outside turn nine with over 100 exotic cars on display, the Firestone Indy Lights Race at 10:15 AM, the Mothers Polishes Exotic Car Parade at 11:30 AM, the Super Truck Race at noon followed by the Izod Indy Car Race at 1:45 PM. followed by the Trans Am Race at 3:45 PM. After that was a Formula Drift demonstration at 3:55 PM and the Pirelli World Challenge Race at 4:30 PM.

Adrian Fernandez and Paul Tracy were Inducted into the Long Beach Motorsports Walk of Fame.

The race was live on NBC Sports with commentators Wally Dallenbach Jr., Townsend Bell, Jon Beekhuis and Robin Miller. Brian Till was the announcer.

Franchitti hopes slump ends after winning pole

HE EDGES HUNTER-REAY IN LONG BEACH

Associated Press

1 minute 7.2379 seconds That bumped body of Funny Car driver Rober

Dario Franchitti had the pole, led the first 28 laps but finished third. Takuma Sato, driving for A.J. Foyt, collected $78,194 for the win followed by Graham Rahal in second and Justin Wilson in third.. Sato passed Hunter Reay for second on lap 23 and took the lead on lap 30. There were five cautions and the race ended under yellow when Tony Kanaan hit the tire barrier in turn one on lap 78. Dixon, hit from behind by Tristan Vautier, Castroneves, Hinchcliffe and Hunter Reay, who hit the tire wall in turn eight, all were involved in incidents which set them back.

Ringo H.W. Chiu Associated Press

TAKUMA SATO races in the lead at the Toyota Grand Prix of Long Beach. Sato, who nearly won the Indianapolis 500 in 2012, won for the first time in 52 starts in the IndyCar series.

Foyt's team gets first win in more than a decade

GRAND PRIX NOTES

Drivers regroup after setbacks

By Jim Peltz

The Toyota Grand Prix of Long Beach was a tough day at the office for several of IndyCar's leading drivers.

Two-time champion **Scott Dixon**, reigning champion **Ryan Hunter-Reay** and **Helio Castroneves** were involved in minor accidents Sunday.

Castroneves and Dixon, however, battled back after their cars were repaired to finish ninth and 10th, respectively, which earned them valuable points.

Castroneves kept the lead in the Izod IndyCar Series standings by eight points over race winner **Takuma Sato** and by 11 points over the third-place Dixon.

Long Beach was the third race of the series' 19-race season.

"I was trying to be careful but got bunched up and broke the front wing," said Castroneves, who drives for Team Penske. "My team did a great job getting me back out."

Dixon spun on the opening lap after being hit from behind by rookie **Tristan Vautier**, "which resulted in a flat tire for us," Dixon said.

Hunter-Reay was running among the leaders early but on Lap 51 of the 80-lap race he drove too fast into Turn 8, a right-hander, and slammed into the track's tire barriers. He finished 24th in the 27-car field.

"Completely my fault, but at that point we were just trying to make a bad day a little better," said Hunter-Reay, a driver for the Andretti Autosport team who won in Long Beach in 2010.

Klaus Graf and Lucas Luhr won the American Le Mans Series Race with Nick Heidfeld and Neel Jani second and Jon Bennett and Colin Braun third. Rutledge Wood won the Toyota Celebrity Race with Brett Davern second and Michael Trucco third. Carlos Munoz won the

Indy Lights Race with Gabby Chaves second and Sage Karam third. Justin Lofton was first in the Super Truck Race with Robby Gordon second and Rob MacCachren third.

Pirelli World Challenge

Race Photos

Long Beach's racy streets

Fast cars and sunshine mix well at the city's Grand Prix

CHRIS ERSKINE
FAN OF THE HOUSE

I've tumbled down another rabbit hole, wound up in Long Beach, which isn't a worst-case scenario, though close. The cars here all need mufflers, the young fans need more clothes. What kind of sordid little event is this?

queen Stephani Costanza remembers taking a lap in a hydroplaning pace car and demanding to be let out.

"I was scared to death," she recalls.

It's worth noting that they don't hold this race on a track. They hold it downtown, by the aquarium and the Hyatt, along the same boulevards you might find yourself some idle summer day before realizing: "YIKES, I'M IN LONG BEACH? GET ME OUTTA HERE!"

tana, after which a champion is crowned based on points.

That's a pretty good SoCal presence: two stops out of 19 total races. After this weekend, off they go to Brazil, then Detroit and into summer, two or three events a month.

It's a world I only pretend to understand — much like the rest of the world. But I do appreciate it. Because the other day, I took a couple of laps around the two-mile trapezoid and am now

shot.)

Or the entertainment areas, where the drinks flow, the music plays. Celebrities prepare to circle the course in their souped-up Scions. Gear heads wander amid the Porsches in the Le Mans paddock.

Along the water, it feels 10 degrees cooler, and you can almost hear again. There are food trucks and miles of merchandise. Tecate beer runs in vast, cold rivers.

Of course, the race is the thing, and it's hard to over-

This could be his year

Castroneves, who has never won an IndyCar series title, leads after two races before Sunday's Grand Prix of Long Beach.

BY JIM PELTZ

Helio Castroneves has won the Indianapolis 500 three times, earned 27 career victories overall and he even placed first in the "Dancing with the Stars" contest one year.

NICK LARAR Getty Images

HELIO CASTRONEVES looks confident before the race in Birmingham, Ala., where he finished third.

Toyota Grand Prix of Long Beach

When: Today-Sunday.
2012 winner: Will Power.

SCHEDULE*
Today: Practice 10:30 a.m. and 2:30 p.m.
Saturday: Practice 10:40 a.m., qualifying 2 p.m. Pro/celebrity race 11:40 a.m.
Sunday: Grand prix 1:40 p.m.

*Other types of race cars practice, qualify and race all three days as well.

Driver wants to leave her mark

De Silvestro hopes to be the first woman to win an IndyCar event since Patrick in 2008.

BY JIM PELTZ

Simona De Silvestro holds out her hands to show there's hardly a sign that they were seriously burned in a racing crash two years ago.

"I wish I had a cool scar

ROBERT LABERGE Getty Images

DARIO FRANCHITTI, four-time IndyCar champion who won the Long Beach Grand Prix in 2009, captured the pole position Saturday with a lap of 105.369 mph.

SCREAMING RUBBER AND STYLISH MOVES

The multimillion-dollar motor sport of drifting brings crowds to their feet and drives demand for go-fast auto parts

BY W.J. HENNIGAN

The tires on Vaughn Gittin Jr.'s Ford Mustang are shrieking as he jerks his car sideways at nearly 90 mph on this serpentine race track in Long Beach.

Walls are closing in on either side, leaving no room for mistakes.

As the next turn nears, Gittin pumps the clutch and yanks a neon green hand brake. The rear wheels lose traction, sending the car into a

What began as an illegal hobby in Southland parking lots has grown into a multimillion-dollar motor sport appealing to a new generation. The Formula Drift series opened its 10th season last weekend on city streets near the Long Beach Arena. On Saturday night, 16 drivers will conduct an encore performance at the Toyota Grand Prix of Long Beach.

"In the 39 years of the Grand Prix, no event has ever taken place at night," said Jim Liaw, president

each paid $27 to $37 for admission, and more are expected at the turnstiles this weekend.

Drifting has nothing to do with racing. There's no checkered flag. It's more like figure skating than speed skating, with a three-judge panel awarding points based on speed, angle of attack and style.

They take points away for going off course, stalling or running into course markers — including walls. And for driving straight: The goal is to control the car as it slides side-

Flashback Heart Attack Performed at the Charity Ball

Super Trucks

Sato, Rahal and Wilson

Toyota Celebrity Race

Formula Drift

A.J. Allmendinger

Takuma Sato

Scott Dixon

Dario Franchitti

Will Power

Sebastian Bourdais

Josef Newgarden

Charlie Kimball

Read the Autoweek article on the attached DVD and watch the video on our you tube channel - www.youtube.com/@racinghistoryproject2607

2014

The 2014 Toyota Grand Prix of Long Beach, race two of the Indy Car season, was held on Aprll 13, 2014.

The Lifestyle Expo featured the Family Fun Zone, BMX exhibitions and "Four Decades of Racing in the Streets", a retrospective of previous races

Friday was practice and qualifying for all. Saturday featured the Toyota Celebrity Race, the Tequila Patron Sports Car Race, Pirelli World Challenge and the Super Truck Challenge.

The Tequila Light Fiesta featured "Kinky" on Friday night'; Saturday nights Rock-N-Roar Concert featured "Paul Rodgers of Bad Company"

The Motegi Super Drift Challenge took place, under the lights on both Friday and Saturday nights. Sunday, race day, had the Indy Lights Race at 11:25 AM and the Indy Car race at 1:00 PM.

Keith Kalkhoven, Dario Francitti and Gerald Forsythe were inducted into the Long Beach Motorsports Walk of Fame.

The race was live on NBC Sports with commentators Paul Tracy, Townsend Bell and Robin Miller. The announcer was Leigh Diffey.

Hunter-Reay wins Long Beach GP pole

Associated Press

Ryan Hunter-Reay, determined to rebound from a disappointing season, is off to a strong start in what he hopes will be another run toward the IndyCar championship.

Hunter-Reay won the pole Sat-

what he has accomplished tl last two weeks in becoming tl youngest Nationwide driver wi multiple victories.

Elliott acknowledged he w still in shock over his first seri win at Texas Motor Speedw when he doubled up at Darlingt

Ryan Hunter Reay had the pole but Mike Conway won after Hunter Reay, leading, made an ill advised pass on leader Josef Newgarden in turn 4 on lap 56, causing a five car pileup that nearly blocked the track completely and, taking out Kanaan, Sato, Hawksworth and Hinchcliffe in addition to himself and Newgardeni. Conway and Will Power were two of those who snuck through the rubble and Conway eventually ended up winning the race just 0.9 seconds ahead of Power, collecting $68,889 for the win. Carlos Munoz was third. Dixon, after an incident with Wilson, ran short of fuel with two laps left, Bourdais hit the tire walls in turn eight twice,

Conway gains late lead, wins race

Dixon runs out of gas while leading with only two laps left, and the Briton takes over.

By Jim Peltz

The first half of the 40th Toyota Grand Prix of Long Beach generally was a stately affair.

There were no major accidents, though Sebastien Bourdais slammed into the tire barriers twice, and Ryan Hunter-Reay maintained the lead with apparent ease.

Scott Pruett and Guillermo Rojas won the Tequila Patron Sports Car Race in a Riley / Ford followed by Ricky and Jordan Taylor in a Dallara Chevrolet and Christian Fittipaldi and Joao Barbosa in a Coyote Chevrolet. Gabby Chaves won the Indy Lights Race with Zach Veach in second and Matthew Brabham in third. Robby Gordon won the Super Truck Race with E.J. Viso second and Sheldon Creed third. Brett Davern won the Toyota Celebrity Race and was the first amateur, followed by Max Theriot and Adrien Brody.. Al Unser Jr. won the professional category.

(Read All The Race Results On the Attached DVD)

Toyota Celebrity Race

Power has been surging in IndyCar

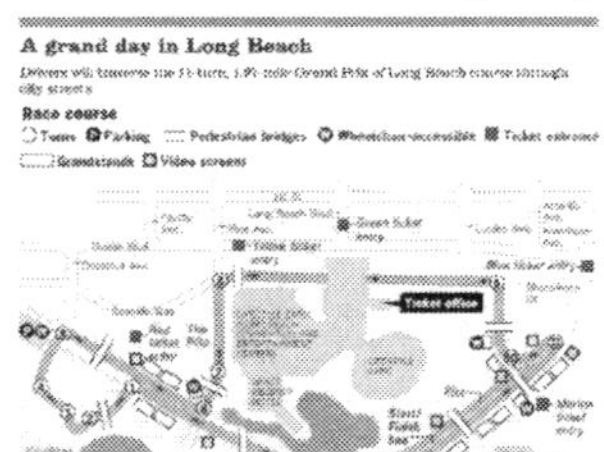

Montoya is fourth in his return

By Jim Peltz

Juan Pablo Montoya celebrated his return to Long Beach after 14 years with a fourth-place finish.

Montoya won the Toyota Grand Prix of Long Beach in 1999 but then left for Formula One racing in 2001 and, six years later, NASCAR stock car racing. He's back in IndyCar with Team Penske.

The Colombian started 16th, was sent to the back of the 23-car field for improperly entering the pits when they were closed but then gained several spots thanks to a multicar crash midway through the race.

"I could have been a little more aggressive at the end, but I really wanted to make sure we didn't get in any trouble and got a decent finish," he said.

Bourdais' troubles

Sebastien Bourdais had high hopes of winning the Long Beach race for the fourth time.

Instead, the French driver twice plowed into the tire barriers in Turn 8 of the 11-turn street course when he was unable to slow enough through the sharp right-hander.

"This one really hurts," said Bourdais, who started third in the Verizon IndyCar Series race and finished 14th. "I had a great car and I just made a mistake."

Wilson, Dixon bump

Near the end of the race, **Justin Wilson** attempted to pass **Scott Dixon** for the lead. Instead, their cars touched and Wilson's went into the wall.

Wilson was none too happy about Dixon's actions. "To be taken out in a way that is so deliberate and blatant, and not even see a penalty for it. I'm a little confused," Wilson said.

"I feel bad for the accident" with Wilson, Dixon said. "I didn't expect him to be there on the outside, so I will go talk to him."

Indy Lights race

Gabby Chaves, 20, won the Long Beach 100, a race in IndyCar's second-tier Indy Lights Series. Pole-sitter **Zach Veach** was second.

james.peltz@latimes.com

This is one 'crown' no athlete strives for

Castroneves still seeking first win

Hunter-Reay criticized after crash

[**IndyCar** from C1]

day.

Mike Conway then inherited the lead and the British driver held on to win his second Long Beach Grand Prix. His first victory on the city's seaside streets came in 2011.

"I can't believe it," said Conway, who started 17th in the 23-car field. "I wasn't sure if Scott was going to pull in [for fuel]. You just have to push as hard as you can all the time."

Australian Will Power, who won the Verizon IndyCar Series' season opener in St. Petersburg, Fla., two weeks ago, finished second.

Rookie Carlos Munoz was third and Juan Pablo Montoya — making his first Long Beach start in 14 years after returning to IndyCar racing — finished fourth.

Conway, 30, is a part-time driver for the team owned by fellow IndyCar racer Ed Carpenter. Conway formerly competed during the entire IndyCar season. But near the end of the 2013 season, while driving for A.J. Foyt Racing, Conway announced he was "not comfortable" racing anymore on high-

Gina Ferazzi Los Angeles Times

EARLY LEADERS Takuma Sato, front, and Ryan Hunter-Reay leave pit row after being knocked out of the race after a multicar crash on Lap 56.

left leg that required months of rehabilitation.

So Carpenter's team hired him this year to drive only on the series' twisty street and road courses, while Carpenter himself

team. He triggered Sunday's big wreck on Lap 56 when he tried to pass Josef Newgarden, who had just exited the pits.

As they entered Turn 4, a right-hander on the 11-turn

Kanaan, defending Long Beach winner Takuma Sato and rookie Jack Hawksworth.

But Conway, Power and Dixon were among those snaking through the wreckage. "You don't get those gifts often in racing, so I'll take it," Power said.

Hunter-Reay's daring move quickly was criticized, even in his own camp, with his team co-owner Michael Andretti saying, "You need to be a little more patient."

Newgarden drives for Sarah Fisher Hartman Racing, and Fisher said on Twitter, "It was our race to win and we got robbed by immaturity. Period."

Hunter-Reay, 33, was less than apologetic after finishing 20th in the 23-car field.

"I'll look at it again, but a racing driver, when he's in the moment and he sees a chance to go for it — I went for it because I want to win the race," he said.

The wreck "got others that didn't deserve to be involved, and that's what I feel really bad about," Hunter-Reay added.

After the race restarted

Ryan Hunter-Reay (28) drives through turns to win the pole position during qualifying for the IndyCar Grand Prix of Long Beach on Saturday. AP

COMP

SPARKS FLY IN INDYCAR

Read the Autoweek article on the attached DVD and watch the video on our You tube channel - www.youtube.com/@racinghistoryproject2607

Tequila Patron IMSA Race

Big Raceday Crowd

Ryan Briscoe

Takuma Sato

Mike Conway

Graham Rahal

Will Power

Charlie Kimball

James Hinchcliffe

Marco Andretti

Conway, Power and Munoz

Autograph Session

IT STARTED IN 1975...
Brian Redman was the first winner, followed by great racing names like Andretti, Lauda, Villeneuve, Unser, Tracy and Franchitti.
APRIL 11-13
THE TRADITION CONTINUES...
40th TOYOTA GRAND PRIX OF LONG BEACH
...featuring Scott Dixon, Helio Castroneves, Juan Pablo Montoya and all the cars and stars of the Verizon IndyCar Series, plus
★Motegi Racing Super Drift Challenge under the lights Fri./Sat. nights!
★Tequila Patrón Sports Car Showcase
★Pro/Celebrity Race
★Pirelli World Challenge
★Robby Gordon's Super Truck Challenge
★Indy Lights
Two free concerts!
Tecate Light Fiesta Friday starring "Kinky" and Saturday Rock-N-Roar featuring "Paul Rodgers of Bad Company".
Plus, 3 days of non-stop family fun with the Lifestyle Expo, Family Fun Zone, BMX and "4 Decades of Racing in the Streets" – a retrospective featuring all of the memories from previous races
DON'T MISS AMERICA'S #1 STREET RACE.
40 YEARS... AND still accelerating!
Get your tickets today!
Visit gplb.com or Call 888-82-SPEED
gplb.com twitter facebook Instagram

2015

TOYOTA
No Speed Limits!
Toyota Grand Prix
of Long Beach

The 2015 Toyota Grand Prix of Long Beach, race three of the Indy Car season, was held on Aprll 19, 2015 in front of a crowd of over 70,000 and a three day attendance of 181,000.

First was the Formula E Race on April 4th. A one day event, the course was shorter than the Indy Car circuit. 20,000 fans took advantage of the free admission

The Lifestyle Expo featured Kid Racers, K1 Speed Go Karts,BMX and skateboard exhibitions and the Indy Car Fan Village.

Friday was practice and qualifying for all and the Fiesta Friday Concert starring "Molotov".

Saturday featured the Toyota Celebrity Race and the Tequila Patron Sports Car Race, Pirelli World Challenge, Speed Energy Super Trucks and the KMC Wheels Super Drift Challenge on Friday and Saturday night plus the Rock 'n Roar Concert with "Vince Neil and Motley Crue".

Sunday, race day, had the Indy Lights Race, then the Indy Car Race at 1:00 PM..

Robby Gordon and Bryan Herta were inducted into the Long Beach Motorsports Walk of Fame.

The race was live on NBC Sports with commentators Paul Tracy, Townsend Bell and Robin Miller. Brian Till was the announcer.

Castroneves earns pole by setting track record

By John Marshall
Associated Press

LONG BEACH, Calif. – An IndyCar mandate forced Chevrolet to ditch its winglets for the Grand

"Every time I go to th race track, I feel like w not only have great spee and a great team — an that's what it's all about Castroneves said after h 42nd pole, fourth on th

Scott Dixon won with pole sitter Helio Castroneves second and Juan Pablo Montoya third. Third place qualifier Dixon beat second fastest qualifier Montoya into turn one at the start, followed Castroneves and took the lead after the first round of pit stops and led the final 34 laps.. There was only one caution period, when Carlos Munoz lost his front wing and unique this year, every car finished. Dixon collected $90,417 for the win

Dixon finally wins Grand Prix of Long Beach

LONG BEACH, Calif. (AP) — Scott Dixon, like nearly every IndyCar driver, loves racing the streets of downtown Long Beach.

But whether it was bad luck or an inability to close out a race, the New Zealander always found victory well beyond his reach.

That changed with a dominating Sunday under the warm California sun.

Dixon passed Helio Castroneves during a mid-race pit stop and stayed out front over the final 47 laps to win the Grand Prix of Long Beach for the first time.

"I guess I finally got it right," Dixon said.

A three-time IndyCar Series champion, Dixon has not had much luck on the 1.968-mile, 11-turn temporary street circuit, finishing fourth in 2010 for his only previous top 10. He led 22 laps last season, but a longshot gamble on fuel came up short.

Dixon qualified third for this year's race, his best starting position at Long Beach. Once the race started, he quickly passed series leader Juan Pablo Montoya for second and went to the lead when Castroneves nearly collided with another car in the pits on lap 33.

Castroneves had one big challenge, but Dixon turned it back and cruised the rest of the 80-lap race to give Chip Ganassi Racing

Grand Prix girls dive for cover as Scott Dixon sprays the crowd after winning the IndyCar Toyota Grand Prix of Long Beach on Sunday in Long Beach, Calif.

HAMILTON WINS BAHRAIN GP

SAKHIR, Bahrain (AP) — Lewis Hamilton's bid for a third Formula One title is gathering momentum after another convincing win in Sunday's Bahrain Grand Prix, while his Mercedes teammate Nico Rosberg's status as his main challenger is under threat from a resurgent Ferrari team.

Hamilton won the race from pole position to strengthen his overall lead in the standings with his third win in four races, and 36th of his career. He is already 27 points ahead of Rosberg and 28 clear of Ferrari's Sebastian Vettel — the only driver to beat Hamilton so far.

"It doesn't matter who it's against, you try and beat everybody out there," Hamilton said, underlining how the F1 title race looks far from being the two-horse contest between Mercedes teammates it was last season. "It's great to be having a fight with the Ferraris."

The British driver started from pole for the first time in the desert race under floodlights and was largely untroubled, finishing ahead of Ferrari's Kimi Raikkonen and Rosberg, whose braking problem with two laps remaining let in Raikkonen for his first podium in two years.

"Ferrari gave us a really good run for our money and we will need to keep pushing as a team," Hamilton said.

Nelson Piquet Jr. won the Formula E Race with Jean Eric Vergne second and Lucas di Grassi third. .Ricky and Jordan Taylor won the IMSA Tequila Patron Race with Joey Hand and Scott Pruett second and Richard Westbrook and Michael Valiante third.. E.J. Viso won the Super Truck Race with Sheldon Creed second and Robby Gordon third. Ed Jones won the Indy Lights Race with Spencer Pigot second and Felix Serralles third.

Electric cars in ePrix ready to charge onto Long Beach streets

Formula E race, one in a global series, will use part of the Toyota Grand Prix route.

By Jim Peltz

Race cars will be back on the streets of Long Beach on Saturday, but earplugs won't be needed.

Formula E, a new series featuring electric-powered used later this month in the venerable Toyota Grand Prix of Long Beach.

Long Beach is the sixth stop on the inaugural 10-race Formula E calendar, a schedule that spans the globe with races in Europe, South America and Asia.

General admission to the Long Beach race is free as Formula E tries to build momentum, especially with younger fans and those intrigued with electric-car technology.

[...] car," Alejandro Agag, Formula E's chief executive, said in an interview.

Formula E initially thought of having racing on the streets of Los Angeles, but "Long Beach, with its fantastic racing history and a track ready to use, was an option that made more sense," Agag said.

The series is sanctioned by the FIA, the governing body of the Formula One racing series. Formula E's cars are roughly similar in [...] holds the traditional Long Beach Grand Prix.

There are 20 Formula E drivers, including former Formula One drivers Nick Heidfeld and Bruno Senna, on 10 teams.

Five different drivers won the first five races. The most recent winner, at a race on the streets of Miami, was French driver Nicolas Prost, son of four-time Formula One champion Alain Prost.

Because of batteries' limited life, each driver has two [...] few years, improved batteries will enable each driver to need only one car per race.

The series also lacks one of racing's alluring signatures: piercingly loud noise. The sound of Formula E's cars has been compared with hearing a supercharged hair dryer, dentist's drill or futuristic Star Wars vehicle whiz past.

Regardless, Formula E recently said two companies backed by media mogul John Malone, Liberty Global [...] quiring a minority ownership stake.

The terms were not disclosed, but Agag said the two firms are now Formula E's largest shareholder. "To have such solid and strong strategic partners, we were really happy with that," he said.

Formula E is a one-day event. After three practice sessions in the morning, qualifying is at noon and the one-hour race over 39 laps starts at 4 p.m.

Drivers rev up for Long Beach

Legendary seaside course will host the IndyCar Series' third race of the season.

Race facts

WILL POWER is the reigning IndyCar champion and a favorite to win Sunday's race in Long Beach.

Dixon seeks his first victory at Long Beach Grand Prix

New Zealander has 35 IndyCar wins in his career, but his best finish here is fourth.

By Jim Peltz

"I've really got to sort my act out at that place," Scott Dixon says.

That place would be the seaside streets of Long [...]

Race facts

- What: Toyota Grand Prix of Long Beach.
- Series: Verizon IndyCar Series.
- Where: 1.97-mile Long Beach street circuit.
- When: Sunday, 1:30 p.m.
- Distance: 80 laps.
- TV: NBC Sports Network.
- Defending winner: Mike [...]

[...] ished 11th last weekend in the Grand Prix of Louisiana.

That's left Dixon a distant 15th in this year's early championship standings; Long Beach is the third race of the 16-race IndyCar schedule.

"It's a rough start," Dixon acknowledged. "But it's the early part of the season. I think we can turn it around."

And Dixon knows how to overcome an early slump.

He also began 2013 poorly [...]

ROBERT LABERGE/GETTY IMAGES

Helio Castroneves (front, right) and Scott Dixon (front, left) fight for the lead early in the Verizon IndyCar Series Toyota Grand Prix of Long Beach on Sunday.

Toyota Celebrity Race

Patrick Stewart was the Grand Marshal

Big Crowds

Vince Neil and Motley Crue

Speed Energy Super Trucks

Tudor Sports Car Series

Josef Newgarden

Juan Pablo Montoya

Sebastian Bourdais

Scott Dixon

Marco Andretti

Charley Kimball

Lica di Filippi

Jack Hawksworth

Tony Kanaan

Sebastian Saavedra

Dixon, Castroneves and Montoya

Scott Dixon wins Grand Prix of Long Beach

Scott Dixon, like nearly every IndyCar driver, loves racing the streets of downtown Long Beach.

But whether it was bad luck or an inability to close out a race, the New Zealander always found victory well beyond his reach.

That changed with a dominating Sunday under the warm California sun.

Dixon passed Helio Castroneves during a mid-race pit stop and stayed out front over the final 47 laps to win the Grand Prix of Long Beach for the first time.

"I guess I finally got it right," Dixon said.

A three-time IndyCar Series champion, Dixon has not had much luck on the 1.968-mile, 11-turn temporary street circuit, finishing fourth in 2010 for his only previous top 10. He led 22 laps last season, but a longshot gamble on fuel came up short.

Dixon qualified third for this year's race, his best starting position at Long Beach. Once the race started, he quickly passed series leader Juan Pablo Montoya for second and went to the lead when Castroneves nearly collided with another car in the pits on lap 33.

Castroneves had one big challenge, but Dixon turned it back and cruised the rest of the 80-lap race to give Chip Ganassi Racing its first victory this season. It was also the 39th of his career, passing Bobby Unser for fifth place on the victory list.

"Scott Dixon has talent and skill and will win races for years and years to come," Unser said. "He is always really fun to watch, for sure."

So have the Team Penske drivers.

Castroneves, who turns 40 next month, had his third straight top-5 finish with a second. Montoya

Read the Autoweek article on the attached DVD and watch the video on our you tube channel - www.youtube.com/@racinghistoryproject2607

Miss Toyota Grand Prix

NO SPEED LIMITS!
41st
TOYOTA GRAND PRIX
APRIL 17—19
of Long Beach
7 FULL-THROTTLE RACES!
★ The turbocharged cars and stars of the Verizon IndyCar Series,
★ Pro/Celebrity Race, ★ Tequila Patrón Sports Car Showcase,
★ KMC Wheels Super Drift Challenge (Fri. and Sat. nights), ★ SPEED Energy Off-Road Series trucks, ★ Pirelli World Challenge ★ Indy Lights.
3 DAYS OF FUN!
The Lifestyle Expo, filled with KidRacers, K1 Speed Go-Karts, BMX and skateboard exhibitions, video games, IndyCar Fan Village and much more.
PLUS!
Two free (with race ticket) concerts - "Fiesta Friday" starring MOLOTOV and Saturday's "Rock-N-Roar Concert" starring VINCE NEIL of Mötley Crüe.
Don't miss Southern California's
Official Spring Brake!
Tickets at GPLB.COM and 888-827-7333

41st
TOYOTA
GRAND PRIX
LONG BEACH

2016

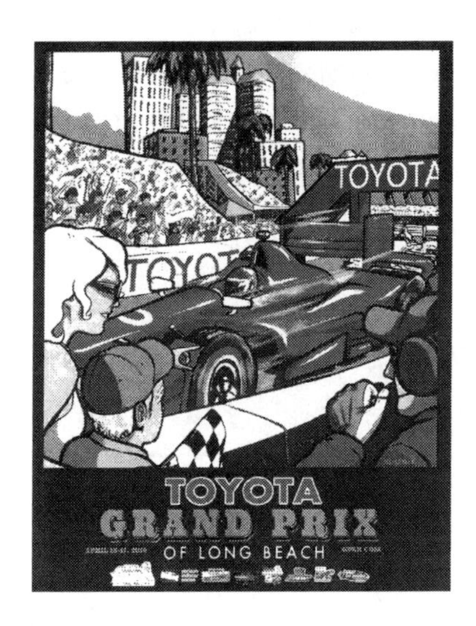

The 2016 Toyota Grand Prix of Long Beach, race three of the Indy Car season, was held on Aprll 17, 2016

The Lifestyle Expo and Family Fun Zone included BMX demonstrations, go karts and driver autograph sessions.

The Formula E Race was held week before was using a shortened version of the Grand Prix circuit. Attendance was 17,000 and admission was free..

Friday was practice and qualifying for all plus the Nortec Collective Concert with "Bostich and Fussible".

Saturday featured the Toyota Celebrity Race and IMSA Bubba Burger Sports Car Race plus Super Trucks and the Nortec Collective Concert with "Cheap Trick".

Sunday, race day, had the Indy Lights Race then the Indy Car Race at 1:00 PM followed by Formula Drift.

Bruce Flanders and Roger Penske were inducted into the Long Beach Motorsports Walk of Fame.

The race was live on NBC Sports with commentators Paul Tracy and Townsend Bell, Robin Miller was the pit reporter.

Castroneves wins pole

LONG BEACH, Calif. — Helio Castroneves made Team Penske 3 for 3 in qualifying this season when he won the pole Saturday for the Grand Prix of Long Beach.

Castroneves turned a lap at 1 minute, 7.1246 seconds to earn the top starting spot for the race today on the temporary street course. Scott Dixon, the winner at Phoenix, qualified second and was

Helio Castroneves had the pole after some confusion as timing and scoring malfunctioned. Castroneves led the first 49 laps with Pagenaud running second and Dixon in third. Pagenaud had a quick pit stop, beating Dixon and Castroneves out of the pits, though questionable as to whether his right side tires were over the blend line on pit exit. He was given a warning and went on to win by 3/10ths of a second over Dixon with Castroneves in third. There were no full course yellows and no DNF's. Pagenaud collected $74,667 for the win.

KIRBY LEE-USA TODAY SPORTS

Simon Pagenaud celebrates after winning the 42nd Toyota Grand Prix of Long Beach on Sunday.

Pagenaud wins 1st race for Penske at Long Beach

ASSOCIATED PRESS

Simon Pagenaud raced to his first victory for Team Penske by holding off Scott Dixon in the caution-free Grand Prix of Long Beach.

It was a controversial win, though, as Dixon and his Chip Ganassi Racing team believed Pagenaud should have been penalized for crossing a blend line as he returned to the track following a pit stop. The Dixon camp interpreted the rule as a clear violation, but IndyCar only gave Pagenaud a warning.

"They told us with the steward system this year there would be no warnings," Ganassi team manager said. When told IndyCar had issued him a warning, the Frenchman said: "I don't care."

Indeed, it didn't matter to him in victory lane for the first time since he joined the Penske group last year. He failed to win a race in 2015, his worst season in IndyCar.

But he's off to a strong start to 2015 with a pair of second-place finishes to start the season and put him atop the points standings headed into Long Beach. He said on the first day of track activity that he knew his first win for Penske was coming, and he was correct.

Dixon was second, while Penske drivers Helio Castroneves and Juan Pablo Montoya finished third and

RACER

Castroneves nabs Long Beach pole

Helio Castroneves took the pole for the 42nd annual Toyota Grand Prix of Long Beach as Chevy swept the top six starting positions.

Castroneves' lap (1:07.1246seconds) earned him his second consecutive pole at the Southern California street circuit and second consecutive pole this season after Arie Luyendyk's longstanding

Sports Illustrated

Simon Pagenaud gets pivotal Long Beach win amid controversy

Three thoughts on Sunday's Toyota Grand Prix race in Long Beach, which was controversially won by Penske's Simon Pagenaud.

Simon Pagenaud faced great expectations when he joined Team Penske prior to the 2015 season. The 31-year-old Frenchman, who lives in the heart of NASCAR country in Charlotte, N.C., had built them himself, winning two races in each of the previous two seasons for Sam Schmidt Racing. It was enough to convince Roger Penske to hire him and build a fourth team around to run the entire Verizon IndyCar season.

Pagenaud actually had a solid first season for Penske, a couple of thirds, a fourth and a fifth, but his 11th rank in the points seemed like a disappointment. It had been speculated that if he didn't win in his second season for The Captain, he might not get a third.

Pagenaud stopped the speculation in its tracks Sunday by winning the non-stop, full-green IndyCar race at the Toyota Grand Prix of Long Beach. It was slightly controversial, but that does not matter.

Alfonso Ribeiro won the 2016 Toyota Pro/Celebrity Race,

ALFONSO RIBEIRO TAKES FINAL CHECKERED FLAG IN BACK-TO-BACKVICTORIES AT 40TH TOYOTA PRO/CELEBRITY RACE

Alfonso Ribeiro won the 2016 Toyota Pro/Celebrity Race, triumphing over the most accomplished field the annual charity racing event has ever seen. The roaring crowd witnessed the stunning performance by Ribeiro, who started the race in the 8th position and successfully avoided any collisions or walls to take the checkered flag.

Ribeiro, now a four-time Toyota Pro/Celebrity winner (1994, 1995 & 2015), is best-known for his role as "Carlton" on "Fresh Prince of Bel-Air." Ribeiro, who is the new host of "America's Funniest Home Videos," had the fastest lap of the day with a time of 1:42.102.

(Read These Articles On the Attached DVD)

Ribeiro wins final Toyota auto race

From wire reports

Alfonso Ribeiro won the 40th and final

Alfonso Ribeiro won the Toyota Celebrity Race and was the first amateur. Second place Max Papis won the professional class with Rod Millen, also a professional in third. This would be the final Toyota Celebrity Race. The Formula E Race was won by Lucas di Grassi with Stephane Sarrazin second and Daniel Abt in third. Jordan and Ricky Taylor won the Bubba Burger IMSA Sports Car Race with Joao Barbosa and Christian Fittipaldi second and Dane Cameron and Eric Curran third. Sheldon Creed won the Saturday Super Truck Race followed by Matthew Brabham and Tyler McQuarrie. On Sunday, Sheldon won again with Robby Gordon second and Matthew Brabham third. Johnny O'Connell won the Pirelli World Challenge Race with Al Parente second.

(See All The Results On The Attached DVD)

World Challenge Race

Formula E has the juice

The electric cars make a whooshing sound instead of the familiar roar, but they're bona fide racing machines piloted by top drivers

BY JAMES F. PELTZ

The Formula E electric cars that will race in Long Beach on Saturday are oddly silent for racing machines, but the event rings loudly for California's fledgling electric-car industry.

The race is a showcase for advancements in electric-car technology. And that's why Faraday Future, an electric-car start-up based in Gardena, signed on as the race's title sponsor and why other electric-vehicle companies are sponsor-

Charged up over Formula E

[Electric, from C1]

said Nick Sampson, Faraday's senior vice president of R&D and engineering.

"Part of growing that market is increasing people's awareness about electric cars and changing people's perceptions," he said. "Many people think of electric cars as being dull and boring, and that mold needs breaking."

Indeed, Formula E cars are not go-karts; they're bona fide race cars piloted by experienced drivers.

The series is sanctioned by the FIA, the same body that sanctions the famed Formula One racing series.

The Formula E drivers include Nelson Piquet Jr., who won the inaugural Formula E race in Long Beach last year, and Mike Conway, a two-time winner of the Toyota Grand Prix of Long Beach, one of the top races in the Verizon IndyCar Series.

For Faraday, the race also allows fans to get a close-up view of the company's concept car, the ex-

NELSON PIQUET JR., center, winner of last year's Formula E race in Long Beach, is joined by runner-up Jean-Eric Vergne, left, and Lucas di Grassi.

on with a range of vehicles" at different price points, Sampson said.

In doing so, Faraday would rival Tesla, whose

gets people "thinking electric cars are cool and gets them familiar with the technology," said Michael Boehm, executive director of

speed by: They would turn to each other with a look of surprise as they heard the cars' whooshing sound rather than the conventional roar

Bubba Burgers IMSA Race

Toyota Celebrity Race

Formula Drift

Helio Castroneves

Max Chilton

Sebastian Bourdais

Conor Daly

Scott Dixon

Jack Hawksworth

James Hinchcliffe

Graham Rahal

Pagenaud, Dixon and Castroneves

Watch the video on our youtube channel - www.youtube.com/@racinghistoryproject2607

Race Photos

HOT CARS. WARM SUN. COOL FUN.

ENJOY 3 DAYS & 2 NIGHTS OF

• 6 Exciting Races • Lifestyle Expo & Family Fun Zone
• Food Trucks • Driver Autograph Sessions
• Concerts
Friday: Nortec Collective presents: Bostich & Fussible
Saturday: Cheap Trick

Kids (12 and under) FREE with paid adult

Tickets: gplb.com or 888-827-7333

2017

The 2017 Toyota Grand Prix of Long Beach, race two of the Indy Car season, was held on April 9, 2017.

The Lifestyle Expo featured more than 150 displays, the Green Power Prix View, the Family Fun Zone with rock climbing walls, racing simulators and race cars for kids 3 to 6 years old Also on display were exotic cars, race cars, a tribute to past racing and driver autograph sessions

Friday was practice and qualifying for all and, in the evening. the Motegi Super Drift Challenge.

Saturday featured the Bubba Burger IMSA Sports Car Race, the Historic Can Am Challenge exhibition, Speed Energy Super Trucks (and again on Sunday) and the Pirelli World Challenge. The Motegi Super Drift Challenge took place again in the evening.

Sunday, race day, had the Can Am Challenge and, the Indy Car Race at 1:30 PM followed by a second Speed Energy Super Truck Race. Emerson Fittipaldi and Tommy Kendall were inducted into the Long Beach Motorsports Walk of Fame.

The race was live on NBC Sports with Townsend Bell and Robin Miller.

MOTOR SPORTS

Castroneves wins pole at Long Beach

The Associated Press

LONG BEACH

Helio Castroneves on Saturday won the pole for the Grand Prix of Long Beach for the third consecutive year.

The Team Penske driver held off five Hondas on the street course for Sunday's race. He barely made the final round of qualifying and needed a track-record lap of 1 minute, 06.22 seconds to secure the pole.

"That's the testament of Team Penske," he said. "We were able to compose ourselves and see what we needed to do. All of a sudden the car came

Castroneves in Turn 11. The penalty cost Pagenaud his two fastest laps and removed him from the rest of qualifying. The reigning IndyCar champion will start last in the 21-car field.

Pagenaud cast some of the blame for the incident on Penske teammate Will Power, whom Pagenaud claimed had slowed on the track and forced him to slow.

"It starts with Power," he said. "On the first lap on reds he was really slow so I caught him and had to abort my lap. All three Penske cars were close, so I was boxed in."

IMSA in Long Beach – Wayne Taylor Racing

DIRK SHADD Tampa Bay (Fla.) Times

Helio Castroneves, right, gave team owner Roger Penske his seventh pole in nine years at the Grand Prix of Long Beach. Castroneves hasn't won at Long Beach since 2001.

Castroneves qualified on the pole for the third consecutive time, had a bad start and lost five spots. Dixon led into turn one followed by Hinchcliffe and Hunter Reay, The first yellow was created by a crash between Power and Kimball. Dixon continued to lead with Hunter Reay second. The next yellow, produced when Rossi's car quit, generated a round of pit stops. Dixon continued to lead with Hunter Reay second until, with five laps to go, Hunter Reay stopped on the track bringing out the final yellow. Hinchcliffe jumped into the lead on the restart and won, 1.8 seconds ahead of Bourdais with Newgarden in third. Hinchcliffe collected $71,875 for the win.

Hinchcliffe is victorious

The IndyCar driver wins the Grand Prix of Long Beach for first victory in two years.

By Alex Shultz

Almost two years to the day he won his last IndyCar race, James Hinchcliffe returned to the podium with an unlikely victory at the 43rd Toyota Grand Prix of Long Beach.

Hinchcliffe had never finished better than third at Long Beach as part of the IndyCar circuit, and Sunday marked his fifth win overall. He started from the fourth

cial event."

In that Indianapolis practice session, a piece of the suspension on Hinchcliffe's car pierced his leg, which caused massive blood loss. He missed the rest of the season, but returned in 2016.

"Everyone knows James' story at Indianapolis — it's an amazing story," Josef Newgarden said. "It's a huge credit to what type of racer he is. He's a die-hard racer through and through, and no one can really question that."

Right off the start, polesitter Helio Castroneves fell back to sixth, unable to hold off such top qualifiers as Scott Dixon and Ryan Hunt-

out the 85 laps (a five-lap increase), but no major crashes. J.R. Hildebrand, who took 11th place, sustained a broken bone in his left hand on the last lap when his car knocked against the car of Mikhail Aleshin.

That there were any slowdowns is a contrast to 2016, in which there were no yellows at Long Beach for the first time since 1989.

"Everyone was kind of on the same strategy last year, which let it play out to not much action, unfortunately," Newgarden said. "This year, it was a lot more mixed up, I feel like there was more passing. I did way more passing than I've done

Newgarden grabbed his best finish at Long Beach in third, followed by Dixon and Simon Pagenaud, the latter of whom took the title here in 2016. Pagenaud's run on Sunday was particularly remarkable, considering he started in the back of the pack after incurring a penalty for making contact with Castroneves during qualifying.

But fifth place will hardly provide Pagenaud with the jubilation Hinchcliffe — born and raised near Toronto — expressed after emerging from his No. 5 Honda.

"When I came into this sport, I felt a huge responsibility, to be honest, to keep

Watch the video on our you tube channel - www.youtube.com/@racinghistoryproject2607

Castroneves on third straight Long Beach pole

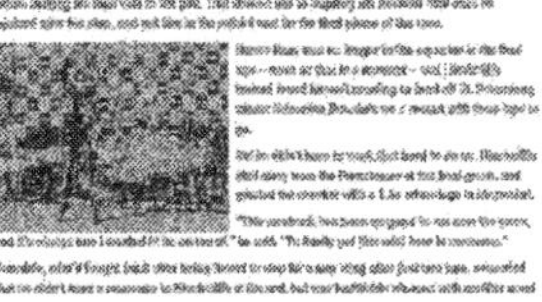

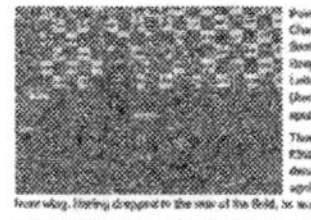

Álvaro Parente captures second Pirelli World Challenge Long Beach GT Event

Read these magazine articles on the attached DVD

Historic Can Am

Helio Castroneves

Josef Newgarden

Carlos Munoz

Al Parente won the World Challenge race with Patrick Long second and Bryan Sellers in third. Ricky and Jordan Taylor won the Bubba Burger IMSA Race with Scott Sharp and Ryan Dalziel second and Jonathan Bomarito and Tristan Nunez third. Matthew Brabham won Saturday's Super Truck Race with Paul Morris second and Gavin Harlien third. On Sunday, Robby Gordon won with Brabham second and Harlien third. Ctaig and Kirk Bennett ran first and second in their Shadows in the Historic Can Am event. Third was Caude Mallette in a Lola T222.

Grand Prix offers chance for a rematch

Pagenaud and Dixon could battle again in Long Beach, but don't count out Bourdais.

By Alex Shultz

Castroneves hoping to end long dry spell

By Alex Shultz

Hello Castroneves is a three-time Indianapolis 500 champion and one of the most successful IndyCar drivers in the sport's history. But his track record at the Toyota Grand Prix of Long Beach is spotty.

Over the last decade, he's had more success to brag about on "Dancing with the Stars" … cause of the power that Hondas seem to be putting out, but that's not going to stop us from working harder. I still want this championship, and hopefully it can be this year."

In the immediate future, Castroneves, whose car has a Chevy engine, is putting his full focus on the Grand Prix of Long Beach, a venue that brings one specific word to his mind.

"Tradition," Castroneves said. "In-

Hildebrand breaks hand; Andretti team has rough day

JIM AYELLO

LONG BEACH, Calif. — Ed Carpenter Racing driver JR Hildebrand broke a bone in his left hand after his No. 21 Chevrolet made contact with Mikhail Aleshin's No. 7 Honda during the final lap of the Grand Prix of Long Beach.

IndyCar announced that its doctors have not cleared Hildebrand to return to competition and that he will be re-evaluated upon his return to Indianapolis.

Hildebrand said the contact occurred as he was trying to … been starting to move over, not a major blocking maneuver but enough to assert his line. He hit the brake a lot earlier than I was expecting and I ended up running into the back of him. In doing so, it ripped the steering wheel from my hand and I ended up tweaking it."

IndyCar penalized Aleshin for failing to yield position.

Hildebrand would finish the race 11th, and considering what happened, he said, it's not a bad result.

"At the end of the day, to come home with an 11th-place finish isn't terrible. It is a bum- … couple of them looked like they were going to finish well, too.

Then ... disaster struck. And it struck hard.

The first victim was Marco Andretti, who suffered what he later called a "sensor issue" that officially ended his day after just 14 laps. Takuma Sato's race came to an end after 78 laps when his car lost power. Sato ended up 18th and Andretti 20th.

But there's more.

It's tough to say who endured the bigger heartbreak, Ryan Hunter-Reay or Alexander Rossi. Both can make an ag- … of tough luck. He was among the leaders who got caught out by the race-deciding yellow flag during the season opener in St. Petersburg.

Hunter-Reay also endured some back luck at St. Petersburg, but he was able to overcome it and finish fourth. That wasn't the case Sunday.

After leading 28 laps, his No. 28 Honda endured an electrical issue — he's pretty sure it was a module that overheated and, because of failsafe, shut the car down. He was running second at the time, on Lap 80 of 85.

"I could see the gap was

Long Beach Grand Prix: Can-Am Challenge roars to life for one day

Joseph DiLoreto tests his 1968 McLaren on a section of track for the upcoming Can-Am Challenge . (P

An ode to one of sports-car racing's glory eras will come to life in a somewhat unlikely way.

Can-Am Challenge Cup cars, those with 1,000-plus horsepower and seemingly louder than a jet engine, will roar down Shoreline Drive for the first time on Saturday as part of the Toyota Grand Prix of Long Beach.

Joe DiLoreto is thrilled to be a part of the proceedings. The retired Long Beach judge will take his

AMG Motorsport teams at Long Beach

Mercedes-AMG to compete in IMSA and PWC races

Special To Highlands News-Sun

AMG Motorsport Customer Racing teams will be competing in both the IMSA WeatherTech SportsCar Championship and Pirelli World Challenge this weekend during the 43rd running of the Grand Prix of Long Beach, today through Sunday. Entries from AMG-Team Riley Motorsports, WeatherTech Racing and SunEnergy1 Racing will race in the GT Daytona (GTD) class in Saturday's 100-minute IMSA WeatherTech Championship race, while CRP Racing, Champ1 and Black Swan Racing will compete in Sunday's 50-minute Pirelli World Challenge race in the GT and GTA classes.

The Grand Prix of Long Beach race weekend hosts the third race on both the 2017 IMSA WeatherTech Championship and Pirelli World Challenge schedules. All competing series at Long Beach use the same 11-turn, 1.968-mile street circuit.

The No. 33 AMG-Team Riley Motorsports AMG GT3 team and drivers earned the first IMSA win for the Mercedes-AMG GT3 in the 12 Hours of Sebring after a successful podium finish in the season-opening Rolex 24 At Daytona. Mercedes-AMG currently leads the Manufacturer Championship standings, the No. 33 AMG-Team Riley Motorsports Mercedes-AMG GT3 leads the Team Championship standings and co-drivers Ben Keating and Jeroen Bleekemolen lead Driver Championship standings in the IMSA WeatherTech SportsCar Championship GTD class.

The No. 75 SunEnergy1 Racing Mercedes-AMG GT3 also saw success at Sebring, qualifying on the pole in the GTD class and ultimately achieving a podium third place finish. Tristan Vautier will co-drive the No. 75 SunEnergy1 Racing Mercedes-AMG GT3 with Boris Said this weekend at Long Beach.

AMG | 10

ALLEN MOODY/STAFF

Ryan Dalziel drives the No. 2 CRP Racing Mercedes-AMG GT3 at Sebring International Raceway.

Bubba Burger IMSA Race

Mikhael Aleshin

Marco Andretti

Sebastian Bourdais

Max Chilton

Conor Daly

Scott Dixon

James Hinchcliffe

Charlie Kimball

43RD TOYOTA
BIGGER... BETTER... FASTER!
GRAND PRIX OF LONG BEACH
APRIL 7-9
• Verizon IndyCars
• IMSA BUBBA burger Sports Car Grand Prix,
• Robby Gordon's Stadium SUPER Trucks,
• Two nights of Motegi Racing Super Drift Challenge,
• Pirelli World Challenge,
• ...and a blast from the past - the Can-Am Challenge - featuring ground-shaking, historic 1000-hp Can-Am cars.
PLUS
TWO EXCITING FREE CONCERTS (w/race ticket)
• Saturday night SMG presents KINGS OF CHAOS starring BILLY IDOL, BILLY GIBBONS of ZZ TOP and CHESTER BENNINGTON of LINKIN PARK,
• Tecate Light Fiesta Friday featuring MODERATTO.
GPLB.COM | (888) 827-7333

Hinchcliffe, Bourdais and Newgarden

2018

The 2018 Toyota Grand Prix of Long Beach, race three of the Indy Car season, was held on Aprll 15, 2018 Three day attendance was over 181,000 making the event the most-attended in the past dozen years. Also announced was that INDYCAR's agreement with the organizing body runs through 2018.

Wednesday featured the Roar to the Shore with motorcycle stunts and race cars on Second Street and the Miss Toyota Grand Prix Contest at the Hotel Maya.. On Thursday, Juan Pablo Montoya and Helio Castroneves were inducted into the Long Beach Motorsports Walk of Fame.

The Lifestyle Expo featured more than 150 displays, the Green Power Prix View, the Family Fun Zone and several driver autograph sessions.

Friday was practice and qualifying for all and an autograph session in the Indy Car paddock and the Fiesta Friday Concert with "Ozomatli".

Saturday featured the Historic Trans Am Race at noon, the Bubba Burger IMSA Sports Car Grand Prix at 1:05 PM, the first Speed Energy Super truck Race at 5:05 PM and at 6:00 PM, "Kings of Chaos" with Billy Idol, Billy Gibbons of ZZ Top and Chester Bebnnington pf Linkin Park headlined the Rock-N-Roar Concert on the plaza stage in front of the convention center. The Motegi Super Drift Challenge started at 7:00 PM.

Sunday, race day, had the Pirelli World Challenge Race 10:00 AM, , the Indy Car Race at 1:40 PM and the second Speed Energy Super Truck Race at 4:05 PM..

The race was live on the Indy Car Radio Network, Sirius XM and NBC Sports with commentators Paul Tracy, Townsend Bell and Robin Miller. The announcer was Leigh Diffey.

Rossi puts Honda on pole at Long Beach

LONG BEACH, Calif. – The way Alexander Rossi sees it, he should be undefeated this season. Instead, he's still looking for his first win of the year.

He wants it to come Sunday on the downtown streets of Long Beach

in hopes of bolstering his racing career. That meant going into major debt last year for a chance to run a few races on the Xfinity Series.

But Preece's investment in himself paid off big time Saturday as the full-

Alexander Rossi had the pole and won the race, leading 71 laps, collecting $71,875. Second was Will Power, 1.24 seconds back with Ed Jones third. Rahal and Pagenaud came together dropping them from contention, though Rahal was able to continue but received a drive through penalty. Kaiser brought out a yellow at the halfway mark. At the restart, Hunter Reay was hit by Sato, had a flat tire and fell back to 19th. Another yellow when De Melo hit the wall had Bourdais and Dixon pitting after the pits had closed. Then King hit Bourdais causing another yellow and forcing Bourdais to pit for a new wing.

Rossi grabs 3rd career IndyCar win

GREG BEACHAM
Associated Press

LONG BEACH – Alexander Rossi spent his early racing career in Europe, and his extended family back home in California hardly ever saw him in person.

With a chance to race an exceptionally fast Honda on the Long Beach streets in front of about 60 friends and relatives Sunday, Rossi showed everybody what they might have missed. He also demonstrated what his Honda looks capable of doing in the rest of this very promising IndyCar season.

Rossi pulled away after a late remember for a very long time for a lot of different reasons."

Rossi comfortably earned the second pole of his three-year IndyCar career during qualifying, and his pace was still there in the race.

Rossi doesn't have many victories yet, but they're all monsters: The 2016 Indianapolis 500 winner and veteran of five Formula One races also won at Watkins Glen late last year.

"I certainly hope I haven't peaked too early with those three," Rossi said of his impressive array of conquered tracks. "I mean, if you're going to hit the wish list, those are the three."

Team Penske Chevrolet back in 14th in the points standings before his standout drive in Long Beach.

"Rossi was just too fast all day," Power said. "He was just really, really good. That was pretty much all we had."

Defending series champion Josef Newgarden finished seventh, and defending Long Beach champion James Hinchcliffe was ninth.

Rossi opened a nine-second lead after the first cycle of pit stops, and he stayed in front after a full-course caution erased the lead with 40 laps to go.

Rossi was losing ground when he pitted again on the 56th lap,

Joao Barbosa and Filipe Albuquerque won the Bubba Burger IMSA Race with Scott Sharp and Ryan Dalziel second and Jordan Taylor and Renger van der Zande third. Gavin Harlien won Saturday's Super Truck Race with Robby Gordon second and Arie Luyendyk Jr. third. On Sunday, Matthew Brabham won with Cole Potts in second and Gavin Harlien third. Daniel Mancinelli won the Pirelli World Challenge Race with Toni Vilander second and Al Parente third.

Long Beach IndyCar: Rossi dominates race to grab championship lead

Alexander Rossi added a street course win to his burgeoning IndyCar résumé at Long Beach, the Andretti Autosport-Honda driver holding off Team Penske-Chevrolet's Will Power in a late-rac shootout.

Rossi led away from pole, while Power made a poor start, then recovered to claim second. But on the outside line, Graham Rahal nudged into Simon Pagenaud and spun the #22 Penske into the wall.

Rahal held the moment to claim third, but went under stewards' investigation and was issued a drive-through penalty. Meanwhile Hunter-Reay

Watch the video on our you tube channel - www.youtube.com/@racinghistoryproject2607 and read the Motorsport.com article on the attached DVD

Tequila Patrón ESM rolls into Long Beach

Team looks for continued success at iconic venue

Special to Highlands News-Sun

LONG BEACH, Cal., – The streets of Long Beach California come alive for this weekend's BUBBA burger SportsCar Grand Prix at Long Beach as the IMSA WeatherTech SportsCar Championship begins the first of three consecutive sprint races on the 2018 calendar.

Twelve Hours of Sebring race winners and Round 2 champions of the Tequila Patrón North American Endurance Cup, Tequila Patrón ESM is ready to get back on track for Round 3 of the IMSA Championship.

First held in the late 1970s, the Grand Prix of Long Beach is a team and driver favorite, transforming the streets of downtown Long Beach into a temporary 1.968-mile, 11 turn-race course. Florida-based Tequila Patrón ESM celebrated success here last year, when Scott Sharp and Ryan Dalziel finished their No. 2 Nissan Onroak DPi prototype in second place. Eager to recommence the on track action, Sebring 12 Hour winners Johannes van Overbeek and Pipo Derani will command the No. 22 Nissan Onroak DPi.

"Long Beach is one of my favorite races of the year," van Overbeek said. "I attended this event as a boy and it's a dream come true to

PATRON | 7

IMSA PHOTO

Tequila Patrón ESM won the 12 Hours of Sebring with its No. 2 car. The team will compete in today's BUBBA burger SportsCar Grand Prix at Long Beach.

PATRON

FROM PAGE 6

participate in it now. This year is running a special meaning as it's also my birthday. I can't think of a better way to celebrate my birthday then to race in Long Beach with Tequila Patrón ESM and race at the circuit. The Brazilian native is in his first season with Acura Team Penske's No. 7 Acura ARX-05 Prototype entry in the WeatherTech SportsCar Championship with co-driver Ricky Taylor.

"It is an honor to see your name on the list, for me, it's just incredible," said Castroneves. "It which includes renditions of the racers' Long Beach-winning cars, as well as their major achievements in motorsports.

Team Joest At Long Beach

Last year's BUBBA burger Sports Car Grand Prix at Long Beach was one of the highlights for

IMSA Bubba Burgers Race

Fiona Murray and Alana Fitzgerald Sang the National Anthem

Kyle Kaiser

Scott Dixon

Ed Jones

Gabby Chaves

Max Chilton

Alexander Rossi

Ryan Hunter Reay

Josef Newgarden

Speed Energy Super Trucks

Long Beach Grand Prix up for grabs

Special to Herald-News Sun

Being a past winner of the Toyota Grand Prix of Long Beach paid off in practice for today's 35th Indy car race on the famous street course in Southern California.

All seven prior race winners entered this year finished in the top 10 on the combined timesheet for the two practice sessions on the 1.968-mile, 11-turn temporary circuit. Scott Dixon, the 2015 Long Beach race winner, was quickest for the day with a lap of 1 minute, 8.4312 seconds (103.582 mph) logged in the first of the 45-minute practices.

Many drivers, Dixon included, were surprised that lap times in the

500," said David Byrd, a principal of Jonathan Byrd's Racing.

Davison's entry is the third for AJ Foyt Racing, paired with full-season drivers Tony Kanaan and Matheus Leist.

The announcement of Davison's entry follows that of Dreyer & Reinbold Racing, which named JR Hildebrand on Wednesday to drive the team's No. 66 Chevrolet with sponsorship from Salesforce. Hildebrand will be paired with Sage Karam in the No. 24 WIX Filters Chevrolet as Dreyer & Reinbold fields two Indy 500 entries for the first time since 2011.

"We've had 87 cars start the Indy 500 over the past close to 20 years now," team owner Dennis Reinbold said.

PRIX

FROM PAGE 2

the 82-foot plunge to the bottom of the Honda Blue Cavern at the aquarium, mingling with the different fish species in the 142,000-gallon tank while answering media questions via an elaborate audio system inside his scuba gear.

"It's like negative G-force down there," Hinchcliffe said, still awash in a smile after emerging from the tank. "It's awesome. In the car, we're getting tossed around and there is a lot of force on the body. Down there, you are as light as a feather. You float around.

"It's one step from being in outer space. It's about as few G's as you can experience on the planet."

As he seasoned [illegible]

"That was awesome!" said Hinchcliffe, the defending Toyota Grand Prix of Long Beach winner. "I'm a fish at heart. I love any chance I get to get in the water.

"That was definitely cooler temperatures than I'm used to diving in, but they kept me nice and prepared and well-dressed for it. Getting to see (the wildlife) that is local to the area is very cool, as I haven't really done much diving on this coast and this side of the country. To be able to have that (special full-face) mask on and be able to talk to everyone that was out there was really cool."

Castroneves, Montoya earn spots on Long Beach Walk of Fame

Team Penske's Helio Castroneves and Juan Pablo Montoya were forever immortalized at the Toyota Grand Prix of [illegible]

SportsCar Championship this weekend for Team Penske on the 1.968-mile temporary street course.

Castroneves raced Indy car 15 times at Long Beach between 1998 and 2017, scoring a victory in 2001 and sitting on the pole three straight times from 2015-17. The 42-year-old Brazilian holds the track lap record of 1 minute, 6.2254 seconds (108.980 mph) set last year in Verizon P1 Award qualifying.

"It's an honor, no question," said Castroneves, the three-time Indianapolis 500 winner who will return in the Verizon IndyCar Series next month for the two races at Indianapolis Motor Speedway. "Just seeing the names of people who are already a part of the Wall of Fame, it just blows my mind away. In 1996, when I came over here to Indy Lights, my first

the CART season championship. The two-time Indianapolis 500 winner drove in five Indy car races at Long Beach.

The 42-year-old from Bogota, Columbia, enjoyed following his boss, team owner Roger Penske, who was added to the Walk of Fame in 2016.

"It's unbelievable," Montoya said. "I came here two years ago to see Roger get inducted. It was huge, huge names here. You don't realize when you race and you win races, you never do it to be here. To get recognized for everything that you've done, it's amazing."

JOE SKIBINSKI/INDYCAR PHOTO

Jordan King and Spencer Pigot with a frosty friend in the Ed Carpenter Racing paddock during the autograph session at Long Beach.

World Challenge

Mario Andretti and the Two Seater

Rossi, Power and Jones

2019

The 2019 Acura Grand Prix of Long Beach, race four of the Indy Car season, was held on Aprll 14, 2019. After Toyota's 44 year run, Acura picked up sponsorship for what is the largest sporting event in Southern California, regularly drawing over 200,000 people for the three day event.

Friday was practice and qualifying for all, an autograph session at 3:30 PM in the paddock and the Fiesta Friday Concert featuring "El Tri" on the plaza stage in front of the convention center at 6:45 PM. The Motegi Super Drift Challenge took place at 7:30 PM.

Saturday featured the IMSA Autograph Session at 9:30 AM in the IMSA paddock followed by the Bubba Burgers IMSA Sports Car Grand Prix Race at 2:06 PM. The first race of the Super Trucks ran at 4:15 PM followed by the Historic GTO Race at 5:00 PM and another Motegi Super Drift Challenge at 6:00 PM.

Sunday, race day, had the Pirelli GT4 Race at 10:00 AM followed by the Mothers Exotic Car Parade. The Indy Car Race started 1:42 PM followed by the second Super Truck Race at 4:05 PM.

The Saturday Night Concert featured "Cold War Kids and Moontower" on the plaza stage in front of the convention center, also at 6:00 PM.

Helio Castroneves and Juan Pablo Montoya were inducted into the Long Beach Motorsports Walk of Fame.

The race was live on NBC Sports with commentators Townsend Bell and Paul Tracy and announcer Leigh Diffey.

Rossi claims pole at energetic Long Beach

JENNA FRYER
Associated Press

LONG BEACH, Calif. – Simon Pagenaud sat down between Will Power and Graham Rahal ready to discuss his best qualifying event of the season.

"I'm still here," Pagenaud defiantly declared.

And with that, personality and panache poured out of IndyCar's top drivers in a highly entertaining

Rossi

The atmosphere is electric at the 45th running of Long Beach, second in prestige only to the Indianapolis 500 on the IndyCar schedule, and the energy around the event has created a spark throughout the field. The race is expected

"The competitive nature of this, like, it drives us all."

Graham Rahal, IndyCar driver

mile, 11-turn clockwise temporary street course has been tight from top to bottom. That, Rahal said, contributed to the light and loose attitude among the drivers after

have to feel a sense of like accomplishment as a team.

"You can see it across all our mechanics, too. Everybody is happy. You make it to the Fast Six, you've really done something. Like this morning, 1.1 seconds across from 1st to 23rd over a street course this long, with all the bumps and curves and this and that, nowhere else in the world will you find racing that compet-

four times he's started an IndyCar race out front.

Scott Dixon qualified second Saturday, his best starting position at Long Beach, followed by the Team Penske trio of Power, Josef Newgarden and Pagenaud. Rahal was sixth for Rahal Letterman Lanigan Racing.

The final qualifying session was split evenly between Honda and Chevrolet, but the Honda drivers

Alexander Rossi was the pole qualifier and race winner for the second straight year, leading 80 laps and collecting $71,875 in prize money.. Second was Josef Newgarden, 20 seconds behind, followed by Scott Dixon in third. Spencer Pigot, Marcus Ericsson, Matheus Leist and Scott Harvey collided in turn three, bringing out a yellow. Ericsson was judged at fault and received a drive through penalty. Herta managed to misjudge his braking point and hit the turn nine wall on lap 50 after staying in the top ten all day. Dixon was handicapped by a botched pit stop that dropped him back to fifth though he recovered to third.

Josef Newgarden

Alexander Rossi

Scott Dixon

Marco Andretti

Rossi cruises to Grand Prix repeat

The driver dominates in the IndyCar Series race in Long Beach a second straight year.

By James F. Peltz

Alexander Rossi turned the Grand Prix of Long Beach into more of a parade than a race.

The Californian won the 45th edition of the famed street race Sunday in even more dominating fashion than he did in winning at Long Beach a year ago.

Rossi captured the IndyCar Series race from the pole position, as he did in 2018, and he led 80 of the race's 85 laps around the 11-turn, 1.97-mile course.

Rossi won by a crushing 20 seconds over second-place Josef Newgarden, the biggest margin of victory at Long Beach since Al Unser Jr. prevailed by 23 seconds in 1995.

"It's an amazing day," Rossi said. "I have a great car and a great crew behind me" on the Andretti Autosport team, he said.

But Rossi also had mixed emotions and was subdued in his victory celebration because his grandfather died

GRAND PRIX OF LONG BEACH REPORT

Herta's return home ends in frustration

By James F. Peltz

Colton Herta of Valencia was hoping to make a second big splash Sunday in his IndyCar Series rookie year, but found the wall instead in the Grand Prix of Long Beach.

Last month, Herta won the series' inaugural race at Circuit of the Americas in Austin, Texas, at age 18, becoming the youngest winner in series history. He turned 19 a week later.

At Long Beach, Herta started 10th and mostly continued running in the top 10 until Lap 50, when he came too hot into the ninth turn of the 11-turn, 1.97-mile street course.

Herta's No. 88 car slammed into the wall, crumpling the left front wing. He managed to get the car to the pits, but his race was over and he finished last in the 23-car field.

"Feel very bad for the team," Herta said on Twitter

No. 9 car on Lap 56, there was a problem with the fuel hose that resulted in Dixon sitting on pit road for a lengthy 18 seconds.

That dropped the Chip Ganassi Racing driver back to fifth, but the five-time IndyCar Series champion managed to finish third. The pit stop "definitely killed us," he said.

Tight spaces

There are spots on the track where the drivers come within an inch or two of the wall, lap after lap. Before the race, **Josef Newgarden** was asked what goes through his mind when he's that close to disaster.

"It's kind of like when you're pulling out of a parking spot and it's tight on both sides and you back up and you start to turn and you're like, 'Man, am I going to miss that car in front of me?' " said Newgarden, who was the 2017 IndyCar Series champion and finished second in Sunday's race.

Grand Prix of Long Beach offers rookie crash course

ARASH MARKAZI

I'm not a very good driver.

It was important for me to warn the fine folks at the Grand Prix of Long Beach of this minor detail before they fitted me for a driver's helmet and put me behind the wheel of a 2019 Acura NSX, a hybrid supercar that has a top speed of 186 mph and goes 0 to 60 in three seconds.

"You'll be fine," said Jim Michaelian, the president and chief executive of the Grand Prix Assn. of Long Beach. "There's no one else

ARASH MARKAZI takes a spin around the Grand Prix of Long Beach course in a 2019 Acura NSX.

ersued by the Clippers for trying to prevent the construction of a competing arena. MSG is headed by James Dolan, who owns the New York Knicks.

"It's not clear to me why it's important for an NBA owner to get involved with another NBA owner in that regard," Ballmer said. "Madison Square Garden would like to have less competition for the music business here in Los Angeles but competition is a good thing; I respect competition. We'll win those lawsuits and build our arena."

The Lakers will finish

The gold standard

Racing venues can learn lessons from Long Beach

LONG BEACH, CALIF.

House of Pistachios was a mom-and-pop startup launched in 1984 by a Southern California family trying to hawk homemade nuts, mints, candies, fruits and chews. As the tiny company grew, the family loaded their wares and traveled up and down the West Coast.

One of the first events they hit was the Grand Prix of Long Beach. The annual four-day downtown

JENNA FRYER

20 seconds on Sunday.

Rossi's performance was flawless, and more than an hour after he took the checkered flag, he used his scooter to weave through the venue past the thousands of party-goers still dancing along to the live music, crowding around televisions to watch the support race that followed the IndyCar event or browsing vendor booths and food trucks.

The days of selling a

Newgarden took a peek into the temporary grandstands during the race itself.

"Who knows if this is true but he said to Scott Dixon and I, 'I was looking around during the race, and no one left. They must have enjoyed it.' Good for him," Rossi said. "I was not spectating on the spectators, but yes, it's an amazing event."

Long Beach has established itself on the SoCal entertainment calendar, thanks to 45 years of date equity. A three-day general admission ticket is $95 (plus one free child admission); a guaranteed seat in the grandstands for all three days is $147 for the best rows and a three-day parking pass at the marina costs $50, though the event is walkable. The

He keeps the grand prix right on track

Jim Michaelian has run the Long Beach event since 2001. This year, he scrambled to find a new sponsor.

JIM MICHAELIAN has been with the Grand Prix Assn. of Long Beach since before its first race in 1975 and assumed the top post in 2001.

BUBBA burger Sports Car Grand Prix coming up

IMSA Weathertech SportsCar Championship heads to Long Beach

If you're at the BUBBA burger Sports Car Grand Prix of Long Beach, or if you're watching on TV, you can't miss that BUBBA burger logo. It's EVERYWHERE.

It's stepped-and-repeated on the signage above the 1.968-mile street circuit's signature "hairpin" turn leading onto the front straightaway, and plenty of other locations around the circuit. It's on the large BUBBA burger boxes that are placed on the roofs of race-winning cars in victory lane.

And if you happen to be anywhere in the Los Angeles area between now and the 2019 BUBBA burger Sports Car Grand Prix on Saturday, April 13, you won't be able to miss it there either.

"This year, specifically, is going to be the biggest year for me on a blitz," said BUBBA burger Vice President of Marketing, Andy Stenson. You can't miss him at the racetrack either. He's the guy wearing the patriotic-themed suit and the flat-billed BUBBA burger hat.

The BUBBA burger Sports Car Grand Prix is set for Saturday, April 13, in Long Beach, Calif.

Another example is distribution of coupons at Long Beach and other IMSA events throughout the year that offer discounts on BUBBA burger products. That also delivered a key learning for Stenson.

"What's interesting is, you can tell that these races are not strictly regional events," he said. "They're really national events, because those coupons will be redeemed – say that we're passing them out at the Long Beach Grand Prix. Well, they may be redeemed in Utah, or they may be redeemed in Ohio, so it's interesting to see, because people are coming from around the country.

"We're seeing redemptions in the Northeast. We're seeing redemptions in the Midwest. We're seeing redemptions in the Pacific Northwest, so we're seeing it everywhere. That's what's been great, too."

(All the race results are on the attached DVD)

Rossi races away to second straight dominant Long Beach triumph

Read the Indy Car article on the attached DVD and watch the video on our you tube channel - www.youtube.com@racinghistoryproject2607

Colton Herta

Pato O'Ward

James Hinchcliffe

Will Power

Max Chilton

Jack Harvey

Ryan Hunter Reay

Zach Veach

Speed Energy Super Trucks

Formula Drift

IMSA Bubba Burger Grand Prix

Rossi, Newgarden and Dixon

2020

Long Beach cancels Grand Prix for now, along with other events in wake of coronavirus concerns

Under orders from city officials, organizers of the Acura Grand Prix of Long Beach on Thursday cancelled next month's 46th edition of the street race amid the growing fears of the Coronavirus.

Just earlier this week, Grand Prix Chief Executive Officer Jim Michaelian said he expected the event to take place with heightened precautions for spectators. But city health officials opted to expand on a state mandate to cancel or postpone events with 250 or more people.

The city order applies to large-scale events including conventions, festivals, parades and sporting events.

"We recognize that this decision affects tens of thousands of residents and visitors, and for some will create immense financial hardship," Mayor Robert Garcia said in a statement. "But our top priority must be the health and well-being of our community and this is absolutely the right thing to do."

The Grand Prix was expected to draw roughly 185,000 people to Downtown on April 17-19.

The Grand Prix Association of Long Beach in a statement said it is in conversations with the city and various race sanctioning bodies to "discuss the viability of rescheduling this event at a later time in the year."

"If that is not possible, then we look forward to presenting the Acura Grand Prix of Long Beach on April 16-18, 2021. Further details about possible refunds or credits will be forthcoming," the organization said.

The decision comes as mounting numbers of conventions and major events, like Coachella and SXSW, have either postponed or cancelled, and the NBA has suspended its season.

Cancelling the annual Grand Prix will likely cost the city millions of dollars in lost hotel rooms, tax revenue and more. A report released in 2018 from Beacon Economics showed that the 2017 Toyota Grand Prix brought $32.4 million for the Long Beach economy and $63.4 million for Southern California.

Jobs could also be impacted. The Grand Prix supports 606 year-round jobs, with 351 of those in Long Beach, the economic impact report found. The event supported labor income for Southern California workers by $24.4 million, including $12.9 million in Long Beach.

It also generates $1.8 million in overall tax revenue, including $700,000 in Long Beach.

2021

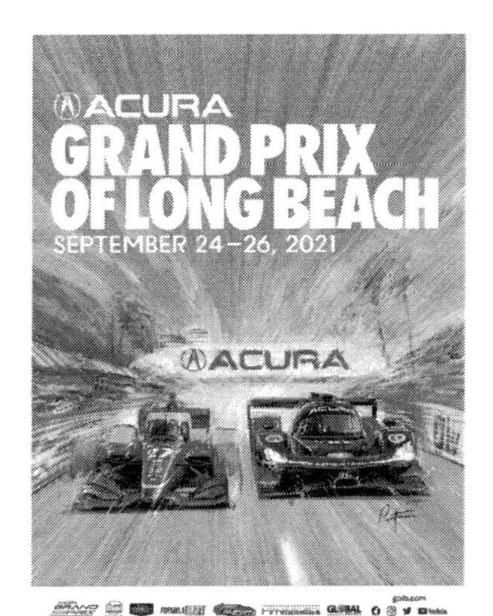

The 2021 Acura Grand Prix of Long Beach, the final race of the NTT Indy Car season, was held on September 26, 2021. The race was originally scheduled for April, 2021 but the date change was required due to the COVID shutdown. Rules required all attendees to wear masks and show proof of vaccination or a negative test to enter. Rapid testing was available at all entrances beginning on the Tuesday before the race.

Friday was practice and qualifying for all and the first of two Formula D Super Drift events.

Saturday featured IMSA Sports Car Grand Prix of Long Beach, a 100 minute race for prototypes and production based cars at 2:00 PM, Global Time Attack on both Saturday and Sunday, a second Formula D Super Drift event and Historic Formula Atlantics. The Speed Energy Super Trucks run on Saturday and again on Sunday. The Saturday night concert featured Vince Neil at 5:30 PM. Sunday, race day, had the Indy Car Race at 12:45 PM.

Long Beach opens up for grand prix return, IndyCar closer

JENNA FRYER
AP Auto Racing Writer

One of the crown jewels of both motorsports and the North American street festival scene opened Thursday for the first time in 17 months after the pandemic snapped a 45-year run for the Grand Prix of Long Beach.

The annual event was one of the longest continuously run street events in auto racing dating to its 1975 opening as a temporary street course through the picturesque Long Beach downtown. The prestige grew during an eight-year run hosting Formula One, and the globetrotting series found it to be a favorable nation or a negative test for COVID-19 taken no earlier than 72 hours before Friday's opening day.

"We're expecting very strong attendance and we want to create the same sense of excitement and energy this event typically carries," Michaelian said Thursday. "We're doing that it in a very safe and secure manner. That's important because when this race concludes (Sunday) this will have been the safest mega-event in California open to anyone to attend."

The 2019 race drew 187,000 spectators over three days, a 1% increase from the year before when Spanish-speaking country since Montoya in 1999. The winner will also be the first champion under 25 since Scott Dixon, who has collected six crowns since that 2003 first title at 23.

O'Ward has said the championship is Palou's to lose, but didn't conceded after falling 35 points behind the leader last weekend at Laguna Seca. O'Ward raced at Long Beach in 2019 (12th-place finish) but Palou saw the circuit for the first time in his life on Thursday.

"I'm going for the win in Long Beach, whatever it takes," O'Ward said. "We still have a shot. I know things can go south real

Willie T. Ribbs and Oriol Servia were inducted into the Long Beach Motorsports Walk of Fame. The race was live on NBC with commentators Paul Tracy and Townsend Bell. Leigh Diffey was the announcer.

Newgarden wins pole in Long Beach

STAFF AND WIRE REPORTS

The IndyCar championship race tightened ahead of the Grand Prix of Long Beach after a messy qualifying session put contenders **Alex Palou** and **Pato O'Ward** in the middle of the

school administrators.
— Eric Sondheimer

Oleksandr Usyk ended **Anthony Joshua's** second reign as world heavyweight champion with a unanimous victory in just his third fight since moving up in weight.

Colton Herta won, leading 43 laps, his second consecutive win, after starting 14th. The win paid $87,143. Pole qualifier Josef Newgarden was second, a half second behind, and led 18 laps. Scott Dixon was third. Alex Palou claimed the Indy Car Championship with his fourth place finish. There were four cautions; on the first lap when Ed Jones rear ended Pato O'ward. Jones received a drive through penalty. Then Ryan Hunter Reay had a flat tire caused by contact with Herta and Bourdais spun and stalled. O'Ward broke a driveshafy and came to a stop on the front straight, Ericsson hit the tire wall in turn one and the final caution occurred when Askew and Daly crashed.

Jerett Brooks won the Saturday Super truck with Matthew Brabham second and Robert Stout third. On Sunday, Robby Gordon won the Super Truck Race followed by Jerett Brooks and Matthew Brabham. Felipe Nasr and Felipe Derani won the IMSA Race with Rwnger van der Zande and Kevin Magnussen second and Tristan Vautier and Loic Duval in third.

Herta pulls ahead in NTT IndyCar Series finale

Sheyanne N Romero
Salinas Californian
USA TODAY NETWORK

LONG BEACH — Californian Colton Herta came in first despite starting 14th on the grid.

Last week, the second-year driver from Valencia dominated the Firestone Grand Prix of Monterey at WeatherTech Raceway Laguna Seca by starting and

can pass here. I'm super happy. This is the biggest race for me outside of (the) Indy (500)."

Newgarden, a two-time series champion, finished second at the 11-turn, 1.9-mile track; Dixon, a six-time champ, placed third.

Finishing fourth was Alex Palou, 24, becoming the first-ever Spaniard to clinch the series points title on the strength of three wins and two poles in

combination over the years. I'm really, really happy for the team. Championships are won by a lot of hard work by a lot of people."

Herta's edgy but crowd-pleasing, come-from-behind victory was his third of the year, sixth of his career and came after the second-generation driver kept his No. 26 Gainbridge Honda out front for 43 of 85 laps.

"Yesterday, I was pretty upset at my-

(See All the Race Results On the Attached DVD)

Palou Takes First Series Title, Herta Earns Win at Long Beach

Alex Palou clinched his first NTT INDYCAR SERIES championship with a composed drive to fourth place Sunday in the Acura Grand Prix of Long Beach, a race won by Colton Herta.

Palou became the first Spaniard to win an INDYCAR SERIES championship and is Chip Ganassi Racing's second consecutive title winner and 14th INDYCAR SERIES champion overall. This is his first season with the powerhouse team in the No. 10 NTT DATA Chip Ganassi Racing Honda, joining CGR after his rookie season in 2020, when he finished 16th in the standings.

"What a race, what a year, what a season," Palou said. "This team is amazing. I'm super proud to be a part of Chip Ganassi Racing, all our partners. I'm super proud to be a champion and for the opportunity these guys gave me.

"Dream completed. Let's get another one now."

Said team owner Chip Ganassi: "Ladies and gentlemen, you're seeing a young man that's going to set a lot of records in this business, and he's already starting. What a great year we've had. I've got to be honest with you; it has surprised me when he came in the door and the job he's done. My hat's off. Congratulations."

Read the Indy Car article on the attached DVD and watch the video on our you tube channel - www.you tube.com@racinghistoryproject2607

Palou becomes first Spaniard to win IndyCar championship

JENNA FRYER
Associated Press

LONG BEACH, Calif. – Smooth and steady, same as he's been all season, Alex Palou cruised to his first IndyCar championship with an easy Sunday drive at the Grand Prix of Long Beach.

The 24-year-old became the first Spaniard to win the IndyCar championship and pulled it off in just his second season in the United States with a fourth-place finish on the temporary

home track – for his second consecutive win and third of the season. Josef Newgarden finished second and Scott Dixon, the six-time and reigning champion, finished third before turning the IndyCar crown over to his Chip Ganassi Racing teammate.

Palou's dream growing up outside of Barcelona was to someday make it to IndyCar and if he was lucky, maybe he could land a ride with Ganassi. He manifested both goals when, as an IndyCar rookie last year, he

ALEX GALLARDO Associated Press

GRANDSTANDING

IndyCars drive into Turn 1 of the Grand Prix of Long Beach, which was won by Colton Herta on his home track for his second consecutive win.

Speed Energy Super Trucks

IMSA

Jimmy Johnson

Pato O'Ward

Scott Mclaughlin

Josef Newgarden

Mario Andretti and the two seater

Historic Formula Atlantic

Herta, Newgarden and Dixon

2022

The 2022 Acura Grand Prix of Long Beach, race three of the Indy Car season, was held on Aprll 10, 2022 with a three day attendance in excess of 187,000.

The Lifestyle Expo featured 270,000 square feet of products and serices on display plus the Green Power Prix View and Family Fun Zone.

Friday was practice and qualifying for all with the Indy Car Autograph Session in the paddock. Friday night. was the first Drift Challenge and, also at 6:30 PM, the concert.

Saturday included more practice, the Indy Car Fast Six Qualifying at 1:00 PM, then the first Historic IMSA GTP race and the IMSA Race at 2:00 PM. The Porsche Carrera Cup Race took place at 4:30 PM, the Drift Challenge again at 6:30 PM and the Saturday Concert starring "Royal Machines" also at 6:30 PM.

Sunday, race day, had the second Historic IMSA GTP Race at 10:00 AM, then the Acura NSX Hot Laps, the Mothers Exotic Car Parade at 11:25 PM and the main event, the Indy Car Race, at 12;45 PM. The second Super Energy Super Truck Race was at 3:30 PM followed by the second Porsche Carrera Cup race at 4:20 PM. Bill Auberlen and Alex Zanardi were inducted into the Long Beach Motorsports Walk of Fame.

The race was live on Sirius XM and on NBC with commentators Townsend Bell and James Hinchcliffe. Leigh Diffey was the announcer.

Herta sets track record, wins Long Beach pole

Jenna Fryer
ASSOCIATED PRESS

LONG BEACH – Colton Herta broke the Long Beach track record in Saturday

the session to a sudden stop.

Herta celebrated as he drove back to pit lane because he thought the session was over. Instead, IndyCar said there were 2 seconds remaining on the clock

Colton Herta qualified on the pole, setting a new lap record. Jimmy Johnson crashed in practice and broke his wrist, requiring a splint on his hand for Saturday's race. Herta led the first 28 laps of the race until passed by Newgarden, who then led 32 laps and won. A yellow brought out by Jimmy Johnson gave Grosjean, who had led 28 laps, a chance and then a second yellow caused by Sato's crash with one lap remaining let Newgarden win under the yellow flag. Alex Palou was third. Newgarden collected $73,750 for the win.

Penske 3-0 after Newgarden win

Jenna Fryer
ASSOCIATED PRESS

LONG BEACH, Calif. – Colton Herta is so good on the downtown streets of Long Beach, and was so strong this weekend, that a second consecutive victory seemed automatic.

Not so fast, scowled Josef Newgarden, who went to sleep the night before the race Sunday stewing over a question he'd been asked after qualifying by a reporter who inquired, "At what point does Colton Herta check out tomorrow?"

"Like what is that? I thought it was such a bizarre question and I went to bed last night and I went, 'You know what? That kid is not checking out. There's just no way,'" Newgarden said.

And so Newgarden kept the pole-sitter in sight when Herta peeled away Sunday in Herta's home race. Herta led the first 28 laps but Newgarden chipped away and used strategy and pit stops to move to the front and win his second consecutive race of the season.

It was his first career victory in 10 starts at the most prestigious street course race in the United States. He was the leader in the 2011 Indy Lights race when he crashed with two laps remaining.

"It's a huge pleasure to finally be able to win around this place," Newgarden said. "It's pretty special to finally get one."

Newgarden led a race-high 32 laps to move Team Penske to 3-0 on the new IndyCar season.

Penske teammate Scott McLaughlin won the opener at St. Petersburg, then Newgarden won at Texas and now Long Beach to bump McLaughlin from the championship lead.

Team Penske last opened an IndyCar season with three consecutive wins in 2012, when the team won the first four races.

Newgarden was challenged over the final 15 laps by Romain Grosjean, the

Josef Newgarden, seen Aug. 21, led a race-high 32 laps to win a second consecutive IndyCar race Sunday on the streets of Long Beach. Newgarden held off Romain Grosjean to seal the victory. JEFF ROBERSON/AP

with Grojean there at the end," said Newgarden.

The victory gave General Motors a sweep in Long Beach; Cadillac went 1-2 on Saturday in the IMSA sports car race and Newgarden won in a Chevrolet.

Grosjean finished second for Andretti Autosport, which was seeking a fourth consecutive win at Long Beach. Andretti drivers Alexander Rossi won in 2018 and 2019, Herta won in 2021 and started from the pole Sunday.

He led 28 laps but had ceded the lead to Newgarden when Herta crashed out of the race with 29 laps remaining. He said he was pushing too hard when he crashed.

"It's just a stupid mistake. We were definitely in that thing," Herta said. "It's unfortunate. I feel really bad. The car was fantastic. Just overdid it a little bit today."

Palou finished third for his 10th podium finish in 19 races since joining Chip Ganassi Racing at the start of last year.

Will Power of Team Penske was fourth and Pato O'Ward salvaged his sloppy start to the season with a fifth-place finish.

The Arrow McLaren SP driver is supposed to be a championship contender but has made mistakes in each of the first three race weekends and is admittedly distracted by his desire for a new contract.

Scott Dixon was sixth for Ganassi, followed by Graham Rahal of Rahal Letterman Lanigan Racing and then Rossi. Helio Castroneves was ninth for Meyer Shank Racing and Kyle Kirkwood was 10th for A.J. Foyt Racing.

Johnson finished 20th to close one of the worst weekends of his career. He broke his hand in Friday's crash, crashed again in Saturday practice, was penalized his two fastest laps in qualifying for interfering with Rahal, and then crashed out of the race Sunday.

"I've had (bad weekends) before," Johnson said. "But it comes with it. I certainly feel bad that I put the team in

Sebastian Bourdais and Renger van der Zande won the IMSA Sports Car Race with Earl Bamber and Alex Lynn second and Tristan Vautier and Richard Westbrook third. Robby Gordon won the first Super Truck Race with Matthew Brabham in second and Max Gordon in third. Max Gordon won the second Super Truck Race with Robby Gordon second and Robert Stout third.

Historic IMSA GTP

IMSA Weathertech Championship Race

Race Photos

(All the race results are on the attached DVD)

Newgarden goes back-to-back with Long Beach win

Josef Newgarden makes it 2 wins in a row and takes over the points lead in the driver standings with Sunday's win at Long Beach. Photo: Chris Owens/IndyCar

After his thrilling win at Texas three weeks ago, Josef Newgarden does it again -- and for the first time ever for him -- in Long Beach

Josef Newgarden became a winner of IndyCar's largest race outside of the Indy 500 Sunday afternoon, holding off Romain Grosjean to take the checkered flag in the Acura Grand Prix of Long Beach.

Starting the race from the second position, Newgarden remained upfront all afternoon before taking the race lead for the final time on Lap 55 of the 85-lap event.

He held on to the lead all the way through the checkered flag despite the best efforts of eventual second-place finisher Romain Grosjean, who attempted to pass Newgarden on the final restart with five laps remaining, but fell short – partly because he was out of push-to-pass boosts that potentially could have took him past the eventual winner.

"It was very close, but not close enough, right?" Grosjean told NBC. "It was fun. We had the right tire strategy, the last portion was going to be great. ... Josef made one mistake but I just couldn't use it (push-to-pass).

"Our Chevy was fast on the straights but I couldn't quite keep up. Very happy to be P2 today, and the

Read the Sports Illustrated article on the attached DVD and watch the video on our you tube channel - www.youtube.com@racinghistoryproject2607

Drivers ready to take it to streets

Series champion Palou one of many who hold Long Beach close to his heart.

By Martin Henderson

The words click off the tip of Alex Palou's tongue as if he's clicking off fast laps on a short oval. "The Indy 500," he says. "Long Beach."

And then there's a long pause.

"I get stuck from there."

That's all you need to know about the importance of the Acura Grand Prix of Long Beach and the prestige it holds in the hearts and minds of IndyCar drivers. In defending series champion Palou's mind, his Mt. Rushmore of race tracks has only two heads.

The 47th running of America's oldest street race takes place Sunday as America's premier open wheel series highlights a weekend of racing that includes the IMSA WeatherTech SportsCar Championship, Robby Gor-

LONG BEACH LINEUP

Lineup for today's Acura Grand Prix of Long Beach (12:30 p.m., Channel 4), with position, car number and qualifying time and speed (mph):

P	#	Driver	C/E/T	Time	Speed
1	26	Colton Herta	D/H/F	1:05.3095	108.480
2	2	Josef Newgarden	D/C/F	1:05.7550	107.745
3	10	Alex Palou	D/H/F	1:05.8667	107.563
4	7	Felix Rosenqvist	D/C/F	1:05.9349	107.451
5	27	Alexander Rossi	D/H/F	1:06.0674	107.236
6	28	Romain Grosjean	D/H/F	No Time	No Speed
7	12	Will Power	D/C/F	1:05.8745	107.550
8	8	Marcus Ericsson	D/H/F	1:05.9548	107.419
9	3	Scott McLaughlin	D/C/F	1:06.0507	107.263
10	60	Simon Pagenaud	D/H/F	1:06.0678	107.235
11	5	Pato O'Ward	D/C/F	1:06.0726	107.228
12	14	Kyle Kirkwood	D/C/F	1:06.2604	106.924
13	15	Graham Rahal	D/H/F	1:06.6896	106.235
14	06	Helio Castroneves	D/H/F	1:06.2467	106.946
15	21	Rinus Veekay	D/C/F	1:06.7049	106.211
16	9	Scott Dixon	D/H/F	1:06.3241	106.821
17	29	Devlin DeFrancesco	D/H/F	1:06.7418	106.152
18	20	Conor Daly	D/C/F	1:06.4489	106.620
19	18	David Malukas	D/H/F	1:06.7925	106.072
20	30	Christian Lundgaard	D/H/F	1:06.5049	106.530
21	45	Jack Harvey	D/H/F	1:06.9708	105.789
22	77	Callum Ilott	D/C/F	1:06.6672	106.271
23	51	Takuma Sato	D/H/F	1:07.1001	105.586
24	4	Dalton Kellett	D/C/F	1:06.7679	106.111
25	48	Jimmie Johnson	D/H/F	1:09.0287	102.636
26	11	Tatiana Calderon	D/C/F	1:07.4789	104.993

Race distance: 85 laps, 167.28 miles. C/E/T: Chassis (D-Dallara), Engine (C-Chevy, H-Honda), Tire (F-Firestone)

and Palou was seventh-fastest, directly behind Herta. Simon Pagenaud of Meyer Shank Racing took honors in 1 minute 07.1991 seconds followed by Andretti Autosport's Alexander Rossi — both with Honda power — and series leader Scott McLaughlin and Josef Newgarden in Team Penske's Chevrolets.

Ganassi's Marcus Ericsson rounded out the top five. The top 13 drivers were separated by less than one second. Several drivers made contact with the wall, including Jimmie Johnson and Pato O'Ward.

Herta broke the Long Beach track record in qualifying on Saturday, and he will start from the pole. Herta turned a lap at 1 minute, 05.3095 seconds to earn his eighth career pole. The previous track record was 1:06.2254 set by Helio Castroneves in 2017.

Palou is currently third in the championship after finishing second and seventh, respectively, on the streets of St. Petersburg and the Texas Motor Speedway oval. Sebastien Bourdais, who was Palou's

Jimmie Johnson injures right wrist in crash

LONG BEACH, Calif. – Jimmie Johnson injured his hand Friday in a crash during practice at the Long Beach Grand Prix. The seven-time NASCAR champion said he needed further evaluation.

Johnson did not remove his hands from the steering wheel – standard practice in open wheel racing but not common in NASCAR – as his car headed into a tire barrier. The force from the collision caused his hands to snap off the wheel and Johnson's in-car camera showed him shaking his right hand.

MATT RANDALL For The Times

JOSEF NEWGARDEN negotiates a curve during the Acura Grand Prix of Long Beach. Newgarden led the last 31 laps as he won his first Long Beach race.

Speed Energy Super Trucks

Will Power

Pato O'Ward

Dalton Kellett

Felix Rosenquist

Colton Herta

Romaine Grosjean

Graham Rahal

Tatiana Calderon

Josef Newgarden

Callum Ilott

Autograph Session

Saturday Night Concert

Newgarden, Grosjean and Palou

2023

The 2023 Acura Grand Prix of Long Beach, race three of the Indy Car season, was held on Aprll 16, 2023 in front of a crowd of over 100,000 with a total attendance for the weekend of 192,000..

Friday was practice and qualifying for all and at 4:55 PM, the Indy Car Autograph Session in the paddock. 6;30 PM was the first Drift Challenge and, also at 6:30 PM, the Fiesta Friday Concert featuring "Boombox Cartel".

Saturday included more practice, the Indy Car Fast Six Qualifying at 12:05 PM, the first Historic Formula One race at 11:20 AM, the IMSA Race at 2:00 PM, the Super Energy Super Trucks at 4:30 PM, the Porsche Carrera Cup Race at 5:15 PM, the Drift Challenge again at 6:30 PM and the "Kings of Chaos" Concert, also at 6:30 PM.

Sunday, race day, had the Historic Formula One race at 10:45 AM, Acura NSX Hot Laps at 11:10 AM, the Mothers Exotic Car parade at 11:25 PM and the main event, the Indy Car Race, at 12;45 PM. The second Super energy Super Truck Race was at 3:30 PM followed by the second Porsche Carrera Cup race at 4:20 PM. Ryan Hunter Reay and James Hinchcliffe were inducted into the Long Beach Motorsports Walk of Fame.

The race was live on Sirius XM and on NBC with commentators Townsend Bell and James Hinchcliffe. Leigh Diffey was the announcer.

KIRKWOOD GRABS 1ST INDYCAR POLE

ASSOCIATED PRESS

Kyle Kirkwood won the first pole of his IndyCar career on Saturday with a fly-

felt it was better for the team."

Marcus Ericsson, winner of the season-opening race, qualified second for Chip

Kyle Kirkwood qualified on the pole, the first of his career and won, leading 53 laps. He collected $64,375 for the win. Romain Grosjean was second and Marcus Ericsson third. O'Ward and Dixon had an incident putting an end to their chance. Newgarden led 27 laps

Kirkwood goes from promising to proven

With guidance from Herta, he wins Grand Prix of Long Beach for his first IndyCar title.

By Steve Henson

Kyle Kirkwood was a phenom on training wheels, perhaps the most successful driver ever along the Road to Indy — what amounts to the bush leagues of racing. Yet after a little more than one year in the IndyCar big leagues, his reputation had turned.

He found trouble. He caused trouble. Somehow, some way, he'd torpedo his own chances and even the chances of others.

Still, his talent was undeniable. So a few weeks ago, the chief strategist for Andretti Autosport, former driver Bryan Herta, switched from advising his son Colton — a rising star in his own right — to work with Kirkwood, the newest member of the Andretti team.

The transformation was immediate, lifting Kirkwood to the pole at the Grand Prix of Long Beach and to the top of the podium Sunday. It was the first IndyCar win for the 24-year-old from Jupiter, Fla.

Jae C. Hong Associated Press

ANDRETTI AUTOSPORT'S newest team member, Kirkwood, en route to his first victory, had a winless rookie season last year.

Kyle Kirkwood

Josef Newgarden

Scott Mclaughlin

Alex Palou

Mathieu Jaminet and Nick Tandy won the GTP Category of IMSA Sports Car Championship Race with Connor de Phillipi and Nick Yelloly in second and Matt Campgell and Felipe Basr in third. Ben Barnicoat and Jack Hawksworth won the GTD category, Patrick Long won the Historic Formula One race driving a Williams. Matthew Brabham won the Speed Energy Super Truck Race with Robby Gordon second and Gavin Harlien third.

(All the race results are on the attached DVD)

Porsche Carrera Cup

Super Energy Super Trucks

IndyCar drama races through new docuseries

A six-episode series on the CW Network will show plenty of Long Beach action.

By Steve Henson

Maybe the surprise is that nobody did it sooner.

Take a sport that stages highly dramatic competition between attractive, purposeful participants in an episodic cadence. Bring on cameras, microphones, scriptwriters, showrunners and producers.

Viewers will stream.

"Formula 1: Drive to Survive," five seasons in and an unabashed Netflix hit, established the model with its revved-up docuseries on the highest class of international auto racing. For proof of the power of unscripted action, view Episode 8 of Season 3 and the horrific crash that snapped driver Romain Grosjean's car in half and enveloped it in flames before he miraculously emerged unscathed 28 seconds later.

"Full Swing" — let's call it Formula 2.0 — shifted the narrative from open-cockpit wonders of aerodynamics that exceed 200 mph to the PGA Tour with an eight-episode series streaming on Netflix.

Set aside the fact that nobody on a golf course places their life in danger so long as an errant shot is accompanied by a throaty "Fore!" Otherwise, racing and golf possess strikingly similar elements that can make for good episodic television: a single-entity organization, reasonably interesting participants, a traveling circus schedule and crescendos of climactic conflict on Sunday afternoons.

The concept also holds true for "Break Point," this year's Netflix documentary series that offers a behind-the-scenes look at the tennis players and tournaments of the ATP Tour and WTA Tour. A second season is in the works.

Next up? "100 Days to Indy," a six-episode series on IndyCar racing directed by Patrick Dimon (College Sports, Inc.) that will pre-

[See Long Beach, D7]

KYLE KIRKWOOD won pole position for the Long Beach race, the first such win of his IndyCar career.

Newgarden to be showcased in new IndyCar docuseries

TEAM PENSKE'S Josef Newgarden (2) will be going for his second consecutive victory at Long Beach.

GRAND PRIX OF LONG BEACH

Walk of Fame Inductees

A film crew had an exciting weekend, filming the new documentary series, "100 Days To Indy" which will include Long Beach.

IMSA Sportscars

Shooting suspect shot by police near Long Beach Grand Prix

ASSOCIATED PRESS

LONG BEACH, Calif. – A man wounded in an exchange of gunfire with police near the weekend Grand Prix of Long Beach racecourse is suspected of being involved in three other shootings in the area, authorities said.

The final shooting occurred Sunday when a patrol officer responding to a report of gunfire shortly before 5 p.m. spotted a man matching the suspect's description, the Long Beach Police Department said in a statement.

The preliminary investigation indicated the suspect fired shots at the officer, the statement said. The man was hospitalized with a wound that was not life-threatening.

Police tweeted an alert to Grand Prix fans to avoid the area and advised that one of the event gates was closed. The course for the prestigious Grand Prix loops through the port city's downtown waterfront streets.

The other shootings occurred earlier Sunday: one shortly after 1 p.m. with no injuries reported, another just before 3 p.m. that wounded a man and another minutes before the gunfire exchange with police with no injuries reported.

Police did not immediately release the name of the suspect.

Witnesses resume ministry at Long Beach Grand Prix

CONTRIBUTED REPORT

Over 180,000 fans will watch in anticipation as world-class racers take to the streets of Downtown Long Beach in the 48th annual Acura Grand Prix, but they will also

there to see smiling faces and share a smile," she said.

Kicking people into high gear is what the Grand Prix is all about expressed Kefflyn Mathews, who has not only been a spectator for many

Local News

CRASH

2023 INDYCAR Acura Grand Prix of Long Beach

Sports Illustrated

Kirkwood dominates for first career IndyCar win at Long Beach

AUTOSPORT

IndyCar Long Beach: Kirkwood scores maiden win to lead Andretti 1-2

Read the Crash, Autosport and Sports Illustrated articles on the attached DVD and watch the video on our you tube channel - www.youtube.com@racinghistoryproject2607

Kyle Kirkwood

Romain Grosjean

Scott Dixon

Colton Herta

Marcus Armstrong

Romaine Grosjean

Autograph Session

Kirkwood, Grosjean and Ericsson

2024

The 2024 Acura Grand Prix of Long Beach, race two of the CART season, was held on Aprll 21, 2024 in front of a crowd totaling 194,000 for the weekend. Beginning the Sunday before was the annual 5K run / walk of the race circuit.

Then on Thursday, Takuma Sato and Katherine Legge were inducted into the Long Beach Motorsports Walk of Fame. There was also Thunder Thursday at the Pike Outlets which featured motocross demonstrations, the Indy Car Pit Stop Competition, a classic car show and beer garden

Open Friday through Sunday, the Lifestyle Expo featured 270,000 square feet of exhibits including Green Technology, the Family Fun Zone with "Lil Lightning Racers", racing simulators, a simulator tournament and bungee jumping. The Fiesta Friday Friday Night Concert with "Ape Drums" took place at the Terrace Plaza at 6:30 PM. On Saturday, the All Star Jam Concert, same time and place, featured Gretchen Wilson, Eddie Montgomery, David Lee Murphy and "Six Wire". Long Beach native Gabriel Iglesias was grand marshal.

Friday was practice and qualifying for all, plus the NTT Indy Car Series Autograph Session in the paddock and the KMC Super Drift Challenge practice and qualifying at 7:00 PM in turn 9, 10 and 11..

Saturday featured the first Historic Indy Car Challenge Race at 11:05 AM, the Tequila Patron IMSA Race at 2:00 PM, the first Super Truck Race at 4;30 PM and the GT America Race at 5:15 PM. The All Star Jam at the Terrace Plaza took place at 6:30 PM while the finals of the Formula D Super Drift Challenge were also at 6:30 PM between turns 9 and 11.

Sunday, race day, had the first Speed Energy Super Truck Race at 7:20 AM, the Indy Lights Race at 8:45 AM followed by the Acura NSX Hot Laps at 11:10 AM, the Mothers Exotic Car Parade at 11:25 AM, and the main event, the NTT Indy Car Race at 12.:45 PM. The second Speed Energy Super Truck Race took place at 3:30 PM followed by the GT America Race at 4:20 PM.. After that, there was a drifting demonstration at 3:55 PM and the Pirelli World Challenge Race at 4:30 PM.

The Indy Car Race and IMSA Race were both live on USA with commentators Calvin Fish and Dillon Welch for IMSA and Townsend Bell, James Hinchcliffe and Welch for the Indy Car race.. Leigh Diffey was the announcer for both.

Rosenqvist career turnaround continues, claims Long Beach pole

Meyer Shank Racing also gets first IndyCar top position

round of qualifying. Alexander Rossi, in a contract year, will start 13th, Pato O'Ward will start 14th and reigning F2 champion Theo Pourchaire will start his IndyCar debut from 22nd. Pourchaire is the

Felix Rosenquist won the pole in Sunday's Fast Six Qualifying session but Scott Dixon, using a fuel stretching strategy, won by less than a second over Colton Herta with Alex Palou third. On lap one, O'Ward ran into Rossi and was given a drive through penalty. Newgarden led for 19 laps and threatened until he got bumped by Herta with ten laps to go. There was only one caution, when Christian Rasmussen spun and hit the wall in turn four on lap 15. That splintered the field with some going to a different fuel strategy. This was Dixon's 57th Indy Car victory and paid $71,250.

(All the race results are on the attached DVD)

Dixon saves fuel to win at Grand Prix of Long Beach

Chip Ganassi Racing driver Scott Dixon celebrates Sunday after winning the Long Beach Grand Prix. The victory was the 57th of his career, moving him 10 wins from A.J. Foyt's all-time IndyCar mark.
GARY A. VASQUEZ/USA TODAY SPORTS

Nathan Brown
The Indianapolis Star
USA TODAY Network

LONG BEACH, Calif. – As only Scott Dixon seemingly can, the Chip Ganassi Racing driver drove a masterclass of a fuel save to the lead in Sunday's Grand Prix of Long Beach, and then held off a late-race blitz from Josef Newgarden to secure win No. 57 of his career – moving the six-time champion 10 wins from A.J. Foyt's all-time IndyCar mark.

To do so, though, Dixon got help from an unpenalized bump from Colton Herta into the back of Newgarden with under 10 laps to go that scrambled the top-5 and helped the CGR veteran maintain his cushion to the end.

"Josef was coming strong, and I was kinda unsure how we could beat him once he got behind us," Dixon said from Victory Lane. "But (team owner Chip Ganassi) said just go for it, and man, I was gonna try.

"This win is way up there, man, and the stress levels were high. Those guys were coming fast and strong. I saw Josef coming and thought, 'Man, this is going to be tough.'"

Here's how he did it:

- **Herta stalls Newgarden with late-race bump.**

With eight laps to go and Newgarden inching within just a couple tenths of the eventual race-winner, runner-up Herta slid into the back of the No. 2 Chevy in the middle of the hairpin leading into the front-straight.

The contact sent Newgarden into anti-stall mode, opening him up to approaches from Herta and Alex Palou, before the Penske driver got back moving again. Newgarden would go on to finish 4th, unable to recover from the bump. On the radio, Newgarden and his strategist (and Team Penske president) Tim Cindric were at a loss for words on how Herta wasn't called for an avoidable contact penalty for the move.

On Lap 1, Arrow McLaren's Pato O'Ward ran into the back of teammate Alexander Rossi around the fountain and was given a drive-thru for an error he admitted later on the radio was his fault and his alone.

"I don't know, it seemed pretty obvious that he just misjudged it and ran into me, and once I got lifted, I went into anti-stall mode and couldn't get going," Newgarden said post-race. "I had to wait for the clutch to disengage and reset and just kinda stalled there for a second.

"I think we've got to be happy with 4th, but I'm just not sure about that Herta deal. I think (IndyCar race control) has to look at that. When you hit somebody, I guess it is what it is, but it's not when I hit somebody, but if we've got to take a 4th, we'll take a 4th."

Sebastian Bourdais and Renger van der Zande won the IMSA Race with Luis Felipe Derani and Jack Aitken second and Dane Cameron and Felipe Nasr third,. Jason Daskalos won the GT America Race with James Sofronas in second and Johnny O'Connell in third. Max Gordon won the Saturday Super Truck Race with Robby Gordon second and Matt Brabham third. On Sunday, Myles Cheek won with Matt Brabham second and Bill Hynes third. Tim DeSilva won the Historic Indy Car Challenge in a 1995 Lola. Jason Daskalos won the GT Anerica SRO3 class, Dan Knox won in GT2 and Isaac Sherman in GT4.

AUTOWEEK

Lexus Celebrates 1989 with First Place in IMSA GTD Class at Long Beach

AUTOWEEK

Long Beach Notes: Colton Herta Takes Blame for Late Miscue in NTT IndyCar Thriller

Read the Autoweek article and the Fan Guide on the attached DVD and watch the video on our you tube channel - www.youtube.com@racinghistoryproject2607

Tom Blomqvist

Scott Dixon

Colton Herta

Marcus Ericsson

Herta takes blame for late-race hit that ruined Newgarden bid

Motor Sports Insider
Nathan Brown
Indianapolis Star
USA TODAY NETWORK

LONG BEACH, Calif. – Colton Herta was waiting for a call that never came, and he couldn't have been happier.

The Andretti Global driver was trying to chase down Scott Dixon with all he had in the closing laps of Sunday's Grand Prix of Long Beach, eventually settling for a runner-up finish. But Herta admitted post-race he wouldn't have been surprised to be called to drop back two spots to 4th-place after his late-race bump into the back of Josef Newgarden that sent the Team Penske driver into anti-stall mode.

The contact, just a subtle nudge from the No. 26 Honda into the back of Newgarden as the pair rolled through the track's famous hairpin corner, came as the driver of the No. 2 Chevy found himself just feet off the back of Dixon – having closed within a half-second in the laps prior as he plotted out what could've been a win-sealing pass. Herta took full blame for the contact that forced Newgarden to slow and drop two spots to his eventual finishing position.

Colton Herta trailed Josef Newgarden into the closing laps of Sunday's Grand Prix of Long Beach, before the pair touched in the famed hairpin, causing the Penske car to stall and the Andretti car to cycle into 2nd place. PROVIDED BY INDYCAR

"I would've been surprised with a drive-thru or something like that, but I would not have been surprised if they moved me behind him. It was borderline," Herta said of what he expected in terms of a ruling in the minutes that followed. "But the problem wasn't the hit. It was when he landed, he went into anti-stall. That's what killed it.

"If he doesn't go into anti-stall, nothing happens, and we all stay in line, and I may have had a chance at the end of the front straight (to pass him). I think you could call to rotate back, but then that would mean (3rd-place finisher Alex Palou) would end up 2nd. So, c'mon. In this sport, we don't need him doing more."

Herta and the defending series champ shared a light chuckle while sitting next to each other on the media center stage. Elsewhere, Newgarden was less than pleased.

See BROWN, Page 3B

Speed Energy Super Trucks

NOS
ENERGY DRINK
INDYCAR
SERIES
United
Rentals
RESORT & CASINO
SAN MANUEL

ACURA
Type

Alexander Rossi

Nolan Siegel

Theo Porchaire

Kyle Kirkwood

Newgarden stripped of win at season-opening race

Cheating discovered 2 months after race

JENNA FRYER
Associated Press

Team Penske suffered a humiliating disqualification Wednesday when reigning Indianapolis 500 winner Josef Newgarden was stripped of his victory in the season-opening race for manipulating his push-to-pass system.

Penske teammate Scott McLaughlin, who finished third in the opener on the downtown streets of St. Petersburg, Florida, also was disqualified. Will Power, who finished fourth at St. Pete, was not penalized but docked 10 points.

Additionally, all three Penske entries were fined $25,000 and forfeited all prize money associated with the race. Power has not been accused of any wrongdoing.

Roger Penske owns the race team, the IndyCar Series and Indianapolis Motor Speedway, host of the Indy 500.

"Very disappointing," Penske said in a text message to The Associated Press. "I am embarrassed."

RYAN SUN, ASSOCIATED PRESS

Team Penske driver Josef Newgarden races during a qualifying session for the IndyCar Grand Prix of Long Beach on April 20.

The reverberations were immediate throughout the paddock.

"I've emulated Roger Penske for many years on and off the track, so today's news is quite a disappointment for me," rival team owner Chip Ganassi told the AP. "This is a blemish on his team, their organization, and the series. Very disappointing as a fellow owner and competitor in the series."

The disqualifications gave the victory to Pato O'Ward, who finished second. It is the first win for McLaren's IndyCar team since 2022.

Although Newgarden is accused of cheating in the March 10 opener, IndyCar said the manipulation wasn't discovered until Sunday morning's warmup in Long Beach, California — nearly two months later.

"The integrity of the IndyCar Series championship is critical to everything we do," IndyCar President Jay Frye said. "While the violation went undetected at St. Petersburg, IndyCar discovered the manipulation during Sunday's warmup in Long Beach and immediately addressed it ensuring all cars were compliant for the Acura Grand Prix of Long Beach.

"Beginning with this week's race at Barber Motorsports Park, new technical inspection procedures will be in place to deter this violation."

A review of the data from the St. Petersburg race showed Team Penske manipulated the overtake system so the three Penske drivers could use push-to-pass on starts and restarts. According to IndyCar rules, the use of the overtake isn't available until the car reaches the alternate start-finish line.

Team Penske President Tim Cindric said in a statement the "push-to-pass software was not removed as it should have been, following recently completed hybrid testing in the Team Penske Indy cars."

"This software allowed for push-to-pass to be deployed during restarts at the St. Petersburg Grand Prix race, when it should not have been permitted," Cindric continued. "The No. 2 car driven by Josef Newgarden and the No. 3 car driven by Scott McLaughlin, both deployed push-to-pass on a restart, which violated IndyCar rules. Team Penske accepts the penalties applied by IndyCar."

Newgarden, a two-time IndyCar champion and reigning Indianapolis 500 winner who is in a contract year with Penske, fell from first in points to 11th with the disqualification. Ganassi driver Scott Dixon, winner Sunday at Long Beach, is now the points leader headed into Sunday's race at Barber Motorsports Park in Alabama.

It's the second cheating offense this season for Team Penske. Joey Logano was fined $10,000 and docked his second-place starting position for a NASCAR race at Atlanta earlier this season because he was wearing an illegal glove during his qualifying run.

Formula Drift

GT America Race

Pit Stop Competition

Driver's Parade

Dixon, Herta and Palou

1975-2025

ACURA GRAND PRIX OF LONG BEACH

APRIL 11-13, 2025

Round Three of the NTT Indy Car Series

Round Three of the IMSA Weathertech Championship

Special This Year:
An exhibition of Formula 5000, Formula One and Indy Cars

Other Books By Dave

IMSA RS		Showroom Stock	
Riverside Vol 1		Riverside Vol 2	
Ascot		Ascot Chronicles	
You Didn't Block Mark Once		Further Faster Longer	

Buy them at www.racinghistoryproject.com, at Autobooks in Burbank or online at 181 Coastal or on Amazon or Ebay

Made in the USA
Middletown, DE
09 November 2024

63973913R00199